AN INTRODUCTION TO ANALYSIS
From Number to Integral

AN INTRODUCTION
TO ANALYSIS
From Number to Integral

Jan Mikusiński
Piotr Mikusiński
University of Central Florida

JOHN WILEY & SONS, INC.

New York   Chichester   Brisbane   Toronto   Singapore

Recognizing the importance of preserving what has been written, it is a policy of John Wiley & Sons, Inc. to have books of enduring value published in the United States printed on acid-free paper, and we exert our best efforts to that end.

ACQUISITIONS EDITOR   Barbara Holland
MARKETING MANAGER   Deborah Riegert
PRODUCTION SUPERVISOR   Katharine Rubin
DESIGNER   Laura Nicholls
MANUFACTURING MANAGER   Andrea Price
COPY EDITING SUPERVISOR   Richard Blander

This book was set in 10/12 Times Roman by Publication Services and printed and bound by Malloy Lithographics. The cover was printed by Malloy Lithographics. Production was handled by Peter Feely at Publication Services. Copyright ©1993 by John Wiley & Sons, Inc.

Library of Congress Cataloging in Publication Data:
Mikusinski, Jan.
    An introduction to analysis : from number to integral / Jan
Mikusinski, Piotr Mikusinski.
      p.   cm.
    Includes index.
    ISBN 0-471-58988-8
    1. Mathematical analysis.   I. Mikusinski, Piotr.   II. Title.
QA300.M544   1993
515–dc20   92-42751
                                                        CIP

Printed in the United States of America

10 9 8 7 6 5 4 3 2 1

Printed and bound by Malloy Lithographing, Inc.

PREFACE

In 1957 my father, Jan Mikusiński, published a book in Poland called *Wstęp do analizy matematycznej* (*Introduction to Mathematical Analysis*). The introduction to that text states, "Students of mathematics and other disciplines that use mathematics (such as chemistry) will find in this book the most interesting and important ideas and problems of mathematical analysis, often introduced in an original way. It will shed a new light on certain concepts and theorems, their role and importance, . . . Graduate students, engineers, and researchers will find that a careful and active reading of it will deepen their understanding of mathematical analysis and improve the precision of their thinking." The book soon become very popular in Poland, and is still in use in a new edition.

Around 1980 my father decided to write a new book on mathematical analysis and invited me to join him. The idea was to build on the experience from the first book while incorporating new ideas in mathematical analysis. The present book is the result of our collaboration. It summarizes a lifetime's experience of one of the world's top researchers and teachers of mathematical analysis.

An Introduction to Analysis: From Number to Integral contains a rigorous exposition of calculus of a single real variable. The core is preceded by two chapters devoted to natural numbers and the construction of real numbers, which provide the foundation of calculus. Chapter 1 introduces the natural numbers. Instead of the standard Peano's axioms, a new, simpler approach is used: Axioms for the natural numbers are formulated in terms of inequalities. In Chapter 2 the construction of the real numbers from the natural numbers, through integers and rational numbers, is presented. A new approach to equivalence relations simplifies the construction of the real numbers. Only the most essential properties of numbers are developed, but enough details are given to show the students the spirit of the process and to give them some idea of what they can legitimately use in a proof. A section on countable sets closes this chapter.

Here an explanation is in order. First, most properties of numbers proved in Chapters 1 and 2 are known to students from elementary school. The purpose of those proofs is not to verify that they are true but to show that it is possible to derive them from the axioms of the natural numbers. Second, in these two chapters, and especially in the exercises, some concepts, such as sequences, which are formally introduced in later chapters, are used. Intuitive understanding of these concepts, developed in the calculus series, is sufficient for the purpose of those chapters.

Chapters 3 and 4 contain essential material about the differentiation side of analysis and the theory of infinite series. These chapters discuss the basic concepts of mathematical analysis: functions, continuity, differentiability, sequences and series of numbers, and sequences and series of functions. The second half of Chapter 4 shows students how to combine differentiation and infinite series to construct important transcendental functions and derive many standard and impressive results of calculus. This gives students some feel for the value of the tools just developed in these two chapters.

Chapters 5 and 6 are short and contain results that ought to look comfortably familiar to students. Chapter 5 contains important theorems on limits of sequences and functions. Chapter 6 is a review of the most important techniques of integration. The stated theorems are carefully proved, and the discussed examples are often more challenging than those usually found in calculus textbooks.

Chapter 7 is probably the most distinctive feature of this book. It contains a direct treatment of the Lebesgue integral, based solely on the concept of absolutely convergent series. This chapter in itself is an equivalent of an entire book. In only a few pages the entire spectrum of the Lebesgue integral is revealed. A student with the usual introductory calculus experience is adequately prepared for the material of this chapter if only given time to adjust to precise arguments. The student who has grown accustomed to the precision of earlier chapters should be well prepared for the challenges of this one. The final sections of this chapter give some applications of the theory. For instance, there is a short treatment of Fourier series.

The exposition in this book is light in tone but rigorous. It is written so as to obtain maximum results with a minimum of work. Theorems are chosen that are very central to the study of analysis yet with an eye on simplicity. The usual practice of the text is not to state the most general theorems but rather to find those that are simultaneously most simple to state and clearly exhibit fundamental themes and concepts. Consistently, the text renders a spare, central core of essential mathematics with just enough applications provided to show why the material is important. For example, some numerical calculations show how theoretical results can drastically simplify calculations.

In general, the level of presentation moves from simple to more complex. We find some very simple arguments in Chapter 1 and some quite sophisticated ones in Chapter 7.

Short dialogs appear throughout the book. They address commonly asked questions and misconceptions. They also give the reader a chance to pause for a moment from the formal language of mathematics.

The text is complemented by a large number of exercises and they are an important element of the text. A student should be able to do almost all of them while reading the book. Some exercises introduce new concepts. Those are not essential for understanding the following sections, but an instructor may decide to assign them as special projects. Routine exercises of the type usually given in a calculus course are given only sparingly. This book is written to follow a standard calculus course, not to replace it. Some problems are not very clearly stated and this is intentional. A precise formulation of the question is an essential part of the problem.

The book is designed as a text for a post-calculus course addressed to students majoring in mathematics or mathematics education. It can also be used for an honors calculus course or a guided study. It will provide students with a solid background for further studies in analysis, deepen their understanding of calculus, and provide sound training in rigorous mathematical proof.

A shorter course may consist of Chapters 3, 4, 5, and 7, with a possible inclusion of Section 2.14 and a possible omission of some proofs in Chapter 7. It is also possible to assign the reading of Chapter 7 to a student with sufficient maturity and mathematical sophistication.

The standard prerequisite for this course is the calculus series. The authors assume that the reader is familiar with the basic set theoretical notation and basic principles of logic. Thus, some students may find it helpful to take a course like Logic and Proof.

November 1992

Piotr Mikusiński
University of Central Florida

ACKNOWLEDGMENTS

It is a pleasure to express my gratitude to those who helped me in various stages of the preparation of this book. The following individuals read my manuscript and made many valuable comments and suggestions: my colleague and friend, Howard Sherwood, University of Central Florida; my students, Karlen Kochar and Michael Mott; the reviewers, who included Thomas K. Boehme, University of California-Santa Barbara; Raimond A. Struble; John A. Synowiec, Indiana University Northwest; and Erwin Kreyszig, Ohio State University. Steven V. Wilkinson, Northern Kentucky University, and my wife Grazyna, generated the illustrations.

I also thank the staff at John Wiley & Sons, especially Barbara Holland, mathematics editor; Lynn Trapp, her assistant; and Richard Blander, senior copy editor, as well as Publication Services, for their assistance in the final stages of this project.

Finally, I thank Michael D. Taylor, University of Central Florida, who not only read the manuscript and shared his comments with me, but was there whenever I needed help.

P.M.

CONTENTS

Chapter 3   Functions   42

Chapter 4   Sequences and Series   91

CHAPTER 1

NATURAL NUMBERS

1.1 AXIOMS

By *natural numbers* (or *positive integers*) we mean $1, 2, 3, \ldots$. We have $1 < 2$, $2 < 3$, and so on. The set of all natural numbers and the sign $<$ are taken, in this book, as *primitive notions;* that is, they are accepted without definition. The sign $<$ is put between two numbers and is read "is less than." The set of all natural numbers is denoted by $\mathbb{N}$.

The set $\mathbb{N}$ and the sign $<$ are not the only possible choice for primitive notions. Instead of the relation $<$ we could take, for instance, addition, $+$, as a primitive notion. However, the relation $<$ seems to be simpler. A small child answers at once such questions as "Which is smaller?" even if she or he cannot count. Similarly, the concept of being *the greatest* or *the biggest* is almost innate in humankind; small children, having a choice of sweets, usually take the biggest one.

Principles that are adopted without proof are called *axioms*. From axioms we can deduce further properties that are not accepted without proof. Those are called *theorems*. The following properties will be adopted without proof.

I Asymmetry Principle If $x, y \in \mathbb{N}$ and x is less than y, then y is not less than x.

A number x is said to be a *least element* of a subset $X \subset \mathbb{N}$, if we have $x < y$ or $x = y$ for every $y \in X$. Now we can declare the second property of natural numbers.

II Minimum Principle Each nonempty subset of $\mathbb{N}$ has a least element.

We say that x is a *greatest element* of X if we have $y < x$ or $y = x$ for every $y \in X$.

III Maximum Principle Each nonempty bounded subset of $\mathbb{N}$ has a greatest element.

It is easy to guess what a bounded set is. However, a strictly logical definition is desirable. A subset X is called *bounded* if there is an element $x \in \mathbb{N}$ such that $y \leq x$ for every $y \in X$. The symbol $y \leq x$ is used equivalently for "$y < x$ or $y = x$."

The relation $y < x$ is often called the *strong inequality* in contrast to the *weak inequality* $y \leq x$. For instance, both inequalities $1 \leq 1$ and $1 \leq 2$ hold, because $1 = 1$ and $1 < 2$. On the other hand, $2 \leq 1$ does not hold, because neither $2 = 1$ nor $2 < 1$ is true. The inequality $y \leq x$ can be read in four different ways: *y is not greater than x; x is greater than or equal to y; x is not less than y;* and *y is less than or equal to x*.

Exercises 1.1

1. Does the set $\mathbb{N}$ with the weak inequality $\leq$ satisfy axioms I–III?
2. Does the set of all integers (with $<$ defined as usual) satisfy axioms I–III?
3. Does the set of all rational numbers (with $<$ defined as usual) satisfy axioms I–III?
4. Does the set of all rational numbers x satisfying $0 \leq x \leq 1$ (with $<$ defined as usual) satisfy axioms I–III?

1.2 FUNDAMENTAL THEOREMS

1.2.1 Theorem (Trichotomy)   *For each pair of natural numbers x and y one and only one of the following relations holds:*

$$x < y \quad or \quad x = y \quad or \quad y < x.$$

Proof.   If $x \neq y$, then by II we have $x < y$ or $y < x$. These two inequalities exclude each other, by I. If $x = y$, then neither of the inequalities can hold, again by I.

We are so accustomed to thinking in terms of the above trichotomy that we hardly feel a need to prove it. It is interesting to observe, however, that this property can be rigorously deduced from the preceding axioms.

1.2.2 Theorem   *The least element in axiom II is unique for each set.*

Proof.   Suppose that there are two such numbers, x and y. Then $x < y$ cannot hold, because then y would not be the least element. Similarly, $y < x$ cannot hold. Therefore, by 1.2.1, $x = y$.

1.2.3 Theorem   *The greatest element in axiom III is unique for each set.*

The proof is similar to that of 1.2.2.

1.2.4 Theorem (Transitivity)   *If $x < y$ and $y < z$, then $x < z$.*

Proof.   The first two inequalities imply that neither y nor z is the least element of the set consisting of the three numbers x, y, z. So x must be the least element, by II, and consequently we have $x \leq z$. If we had $x = z$, then we would have $y < x$, which contradicts $x < y$, by I. Therefore $x < z$.

Exercises 1.2

1. Prove 1.2.3.
2. Is it possible to prove 1.2.4 using III instead of II?
3. Let X be a nonempty bounded set of natural numbers. Prove that, if the least element of X and the greatest element of X are equal, then X has exactly one element.

1.3 ON THE EMPTY SET

In axioms II and III we have the expression "nonempty subset of $\mathbb{N}$," which means that there is at least one element in the set.

—A set is a collection of objects. Therefore, if there are no objects, there is no collection and, consequently, there is no set. Thus the expression "empty set" is nonsensical.

—That is true. But traditionally one uses that expression for a set without elements.

—What for?

—To avoid exceptions. For instance, the intersection of two sets would lose its sense if the sets were disjoint. On the other hand, instead of using the jargon "the intersection of the sets X and Y is an empty set" one may simply say that X and Y have no elements in common, or that they are disjoint.

Should we consider the empty set bounded? Intuitively it seems strange to say that it is unbounded, that is, not bounded. On the other hand, if we follow exactly the definition of a bounded set, we cannot claim that the empty set is bounded. Note that from axioms I–III it does not follow that $\mathbb{N}$ is not empty. If $\mathbb{N}$ were empty, then any subset of $\mathbb{N}$ would be unbounded. This sounds like nonsense, but it is a logical consequence of the definition. In order to avoid confusion, we shall use the word *infinite* for each set of natural numbers that is not empty and not bounded.

Exercises 1.3

1. Does the empty set satisfy axioms I–III?

1.4 INFINITY PRINCIPLE

Properties of natural numbers described in Theorems 1.2.1–1.2.4 follow from axioms I, II, and III. Thus, every set satisfying I, II, and III possesses those properties. It is easy to give examples of such sets that are finite. For instance, the set $\{a, b, c\}$ consisting of elements a, b, and c such that $a < b < c$ satisfies I, II, and III. On the other hand, the existence of infinite sets is not obvious at all. It even somewhat contradicts our intuition. However, in order to obtain all the interesting properties of natural numbers, we now adopt the following axiom for use throughout the rest of this book:

IV Infinity Principle   The set $\mathbb{N}$ is infinite.

In other words, we assume that the set of all natural numbers is not empty and not bounded.

Property IV cannot be deduced from axioms I–III because there are sets satisfying I, II, and III, but not IV. We say that axiom IV is *independent* of axioms I–III. Axiom IV plays the central role in the theory of natural numbers and even in the theory of real and complex analysis. You may be surprised to learn that axioms I–IV fully characterize the set of natural numbers. One can also say that the set of natural numbers is defined by axioms I–IV.

Denote by 1 the least element of $\mathbb{N}$. If $x \in \mathbb{N}$, then the least element of the set of all natural numbers greater than x is denoted by $x + 1$. The existence and uniqueness of such an element follow from II, IV, and 1.2.2.

1.4.1 Theorem (Induction Principle)   *Let X be a subset of $\mathbb{N}$ such that*

(a) $1 \in X$;

(b)   *If $x \in X$, then $x + 1 \in X$.*

Then $X = \mathbb{N}$.

In other words, *no proper subset of $\mathbb{N}$ satisfies* (a) *and* (b).

Proof.   Suppose that, on the contrary, there are elements in $\mathbb{N}$ that do not belong to X, and let y be the least of them. The existence of y follows from II. Let x be the greatest element of X such that $x < y$. There is such an element by (a) and III. In view of (b), we have $x + 1 \in X$. By 1.2.1, one of the relations

$$x + 1 < y \quad \text{or} \quad x + 1 = y \quad \text{or} \quad y < x + 1$$

holds. However, the first relation is excluded by the assumption that x is the greatest element of X such that $x < y$. The second relation drops out because $x + 1 \in X$ and $y \notin X$. There remains the third relation, which implies

$$x < y < x + 1.$$

But this contradicts the definition of $x + 1$.

Exercises 1.4

1. Prove that the axioms I–IV are independent. To do that, you have to find a set X and define a relation $<$ in X such that axioms II, III, and IV are satisfied but I is not. Similarly, you have to find examples showing that II does not follow from I, III, and IV, and that III does not follow from I, II, and IV, and finally that IV does not follow from I, II, and III.

2. Does the set of all even numbers $\{2, 4, 6, \ldots\}$ satisfy axioms I–IV? Does it contradict the claim that "axioms I–IV fully characterize the set of natural numbers"?

3. Prove the following version of the Induction Principle:

   Let n be a natural number and let X be a set of natural numbers such that

   (a) *$n \in X$;*
   (b) *If $x \in X$, then $x + 1 \in X$.*

   Then every natural number greater than n belongs to X.

4. Show that the following version of the Induction Principle can be proved without the Infinity Principle:

   Let X be a subset of $\mathbb{N}$ such that

   (a) *$1 \in X$.*
   (b) *If $x \in X$ and x is not the greatest element of $\mathbb{N}$, then $x + 1 \in X$.*

   Then $X = \mathbb{N}$.

1.5 ADDITION OF NATURAL NUMBERS

Addition is defined by induction as follows:

$$(+) \quad x + 1 = \text{the least natural number greater than } x;$$
$$(++) \quad x + (y + 1) = (x + y) + 1.$$

The result of addition of two numbers x and y is called the *sum of x and y*. Part $(+)$ tells us how to add 1 to a natural number. Part $(++)$ tells us how to add the number $y + 1$ if we already know how to add y. We shall prove, by the Induction Principle, that for any natural numbers x and y the sum $x + y$ is uniquely defined and that

1.5.1 $x < x + y$.

Denote by X the set of all natural numbers z for which the sum $x + z$ is uniquely defined for every natural number x and for which 1.5.1 holds. We have to prove that X contains all natural numbers. By $(+)$ and 1.2.2, we have $1 \in X$. (Theorem 1.2.2 is used here to ensure the uniqueness of $x + 1$ for each fixed x). It remains to show that if $z \in X$, then $z + 1 \in X$. Relation $(++)$ tells us what $x + (z + 1)$, is, and $(+)$ ensures the uniqueness of the symbol. Moreover, since $x < x + z$ is assumed to hold, it follows that

$$x < x + z < (x + z) + 1 = x + (z + 1),$$

by 1.2.4, $(+)$, and $(++)$. Hence $z + 1 \in X$.

It is also worthwhile to note that

1.5.2   *If $x < y$, then $x + 1 \le y$.*

which follows directly from $(+)$.

Addition has the following properties:

1.5.3 $x + (y + z) = (x + y) + z$ *(associativity).*

1.5.4 $x + y = y + x$ *(commutativity).*

1.5.5   *If $x + z = y + z$, then $x = y$ (cancellation law).*

Proof of 1.5.3.   Denote by X the set of all natural numbers z such that 1.5.3 holds for all numbers $x, y \in \mathbb{N}$. By $(++)$ we have $1 \in Z$. Moreover, if $z \in X$, then

$$
\begin{aligned}
x + (y + (z + 1)) &= x + ((y + z) + 1) \\
&= (x + (y + z)) + 1 = ((x + y) + z) + 1 \\
&= (x + y) + (z + 1),
\end{aligned}
$$

which implies that $z + 1 \in X$. By the Induction Principle, $X = \mathbb{N}$, which implies that property 1.5.3 holds for all natural numbers x, y, z.

Proof of 1.5.4.   We are going to proceed in two steps. First we prove that

(1) $1 + y = y + 1$

holds for all natural numbers y. Denote by Y the set of all y for which (1) holds. Evidently, $1 \in Y$, because $1 + 1 = 1 + 1$. Assuming that $y \in Y$, we obtain, using $(++)$,

$$1 + (y + 1) = (1 + y) + 1 = (y + 1) + 1.$$

This proves, by the Induction Principle, that (1) holds for all natural numbers y.

In the second step we denote by X the set of all $x \in \mathbb{N}$ such that 1.5.4 holds for x and each natural number y. By what we have just proved, $1 \in X$. If $x \in X$, then by 1.5.3 and (1) we get

$$(x + 1) + y = x + (1 + y) = x + (y + 1) = (x + y) + 1 = (y + x) + 1 = y + (x + 1).$$

Again applying the Induction Principle, we infer that 1.5.4 holds for all natural numbers x, y.

Proof of 1.5.5.   Denote by X the set of all z for which 1.5.5 holds. We first show that $1 \in X$. Assume $x + 1 = y + 1$. By 1.2.1 we have $x < y$ or $x = y$

or $y < x$. The first inequality implies $x < y < x+1$, which is impossible. Similarly, the inequality $y < x$ is impossible, so we must have $x = y$. In the second step we assume that $z \in X$ and show that $z + 1 \in X$. In fact, if $x + (z + 1) = y + (z + 1)$, then by 1.5.3 and 1.5.4 we have $(x + 1) + z = (y + 1) + z$, and hence, by the assumption that $z \in X$, we get $x + 1 = y + 1$. Taking into account the first step of this proof, we conclude that $z + 1 \in X$.

A further important property of natural numbers is the following:

1.5.6   *If x and y are natural numbers such that $x < y$, then there exists a unique natural number z such that $x + z = y$.*

Proof.   Denote by X the set of all natural numbers u such that $x + u \leq y$. This set is not empty, because it contains the number 1, by 1.5.2. Moreover, it is bounded by y because $u < u + x = x + u$, which implies $u < y$. Let z be the greatest number in X. Then, by the Trichotomy (1.2.1), $x + z < y$ or $x + z = y$ or $y < x + z$. The first inequality is excluded by 1.5.2, because otherwise we would have $x + z + 1 \leq y$, contrary to the definition of z. The last inequality is also excluded, because $z \in X$ and thus $x + z \leq y$. Consequently, $x + z = y$.

Suppose now $x + z_1 = y$ and $x + z_2 = y$. Then $x + z_1 = x + z_2$, and hence, by 1.5.5, $z_1 = z_2$. This proves the uniqueness of z.

The solution z of the equation $x + z = y$ is called the *difference of y and x* and is denoted by $y - x$. Theorem 1.5.6 says that the difference $y - x$ exists in the domain of natural numbers if $x < y$.

—This fact is known to every child in elementary school and is at least as evident as axioms I–IV. Why do we need to prove such an obvious fact?

—We do not at all. We can accept it without proof.

—Why then do we prove it?

—For fun. Just like solving a crossword or completing a picture puzzle. Some people find fun and pleasure in deducing new facts from some quite dissimilar facts.

—I do not see any fun in it.

—Besides fun there is another reason to play with axioms, theorems, and proofs. This reveals the very method of mathematics, its core and spirit. Although we begin with childishly easy arguments, we gradually pass to more difficult ones. It is surprising that, starting from the mere axioms I–IV, one can build the noble edifice of calculus and real and complex analysis. This edifice, supported by four tiny columns, turns out to be immovably strong and absolutely precise at every stage.

Exercises 1.5

1. Prove that $(a + b) + (c + d) = (a + c) + (b + d)$.
2. Prove that $((a + b) + c) + d = a + (b + (c + d))$.
3. Prove that if $a < b$ and $c < d$ then $a + c < b + d$.

1.6 MULTIPLICATION OF NATURAL NUMBERS

The result of *multiplication* of two natural numbers x and y is called the *product of x and y* and is denoted by xy (or sometimes by $x \cdot y$). The product of x and y is defined by induction as follows:

$(\cdot) \qquad x \cdot 1 = x;$

$(\cdot\cdot) \qquad x(y + 1) = (xy) + x.$

Let X be the set of all natural numbers y such that the product xy is uniquely defined for every natural number x. By $(\cdot)$ we see that $1 \in X$. Consider some y such that $y \in X$. Then xy is uniquely defined and so is $(xy) + x$ because of the uniqueness of addition. Thus, by $(\cdot\cdot)$, the product $x(y + 1)$ is uniquely defined, which proves that $y + 1 \in X$. By the Induction Principle the set X contains all natural numbers, so the product xy is uniquely defined for all pairs of natural numbers.

To simplify the notation, we always assume that, in an algebraic expression with both addition and multiplication and without parenthesis, the multiplication is performed first. For instance, $(xy) + z$ can be written without parenthesis: $xy + z$.

Multiplication has the following properties:

1.6.1 $\qquad (x + y)z = xz + yz$ (*distributivity*).

1.6.2 $\qquad x(yz) = (xy)z$ (*associativity*).

1.6.3 $\qquad xy = yx$ (*commutativity*).

1.6.4 $\qquad$ *If* $xz = yz$, *then* $x = y$ (*cancellation law*).

Proof of 1.6.1. Denote by X the set of all $z \in \mathbb{N}$ such that 1.6.1 holds for z with each pair of natural numbers x, y. We have $1 \in X$, by $(\cdot)$. If $z \in X$, then

$$(x + y)(z + 1) = (x + y)z + (x + y) = (xz + yz) + (x + y)$$
$$= (xz + x) + (yz + y) = x(z + 1) + y(z + 1)$$

(by $(\cdot\cdot)$, the assumption that $z \in X$, Exercise 1.5.1, and again $(\cdot\cdot)$). Hence, by the Induction Principle, X contains all natural numbers.

Proof of 1.6.2. If for some $z \in \mathbb{N}$ the equality 1.6.2 holds for all $x, y \in \mathbb{N}$, then in view of

$$x(y(z + 1)) = x(yz + y) = x(yz) + xy = (xy)z + xy = (xy)(z + 1)$$

(justified by $(\cdot\cdot)$, 1.6.1, assumption on z, and again $(\cdot\cdot)$) we obtain equality 1.6.2 with $z + 1$ instead of z. But the equality holds for $z = 1$, by $(\cdot)$. By the Induction Principle this proves 1.6.2 for all natural numbers.

Proof of 1.6.3. We first prove that

(1) $$1 \cdot x = x \cdot 1$$

Denote by X the set of all natural numbers x for which (1) holds. Obviously, $1 \in X$. Moreover, if $x \in X$, then

$$1 \cdot (x + 1) = 1 \cdot x + 1 = x \cdot 1 + 1 = x + 1 = (x + 1) \cdot 1$$

(by ($\cdot\cdot$), (1), ($\cdot$), and again ($\cdot$)). Thus $x + 1 \in X$. By the Induction Principle it follows that X contains all natural numbers.

Assume now that 1.6.3 holds for some $y \in \mathbb{N}$ and all $x \in \mathbb{N}$. Then

$$x(y + 1) = xy + x = yx + x = yx + 1 \cdot x = (y + 1)x$$

(by ($\cdot\cdot$), assumption on y, ($\cdot$), and 1.6.1). Therefore 1.6.3 continues to hold when y is replaced by $y + 1$.

Proof of 1.6.4.   Suppose $xz = yz$ and $x < y$. Then by 1.5.6 there exists a natural number u such that $x + u = y$. Since, by 1.6.1,

$$xz + uz = (x + u)z = yz = xz,$$

we obtain the nonsensical result $xz < xz$, by 1.5.1. The inequality $x < y$ is thus excluded. Similarly we exclude $y < x$, and hence the only remaining possibility is $x = y$.

Exercises 1.6

1. Prove that $(a + b)(c + d) = ac + ad + bc + bd$.
2. Prove that $a < b$ implies $ac < bc$.
3. Prove that $a < b$ and $c < d$ implies $ac < bd$.
4. Prove that $a < ab$ whenever $1 < b$.

1.7 ADDITION TABLE AND MULTIPLICATION TABLE

We adopt the following notation:

$$
\begin{array}{lll}
2 = 1 + 1, & 3 = 2 + 1, & 4 = 3 + 1, \\
5 = 4 + 1, & 6 = 5 + 1, & 7 = 6 + 1, \\
8 = 7 + 1, & 9 = 8 + 1, & 10 = 9 + 1.
\end{array}
$$

(1)

The numbers $1, 2, 3, 4, 5, 6, 7, 8$, and 9 are called *digits*. The symbol 0 (which does not represent a natural number) is also a digit. The number 10 is a two-digit number.

The multiplication of two digits will be always denoted by placing a point between them (e.g., $2 \cdot 3$ or $7 \cdot 5$). The symbol 75 will denote $7 \cdot 10 + 5$. Generally, if $\mathbb{A}$ and $\mathbb{B}$ denote any two nonzero digits, we assume that

(2) $$\mathbb{AB} = \mathbb{A} \cdot 10 + \mathbb{B}.$$

If $\mathbb{B} = 0$, then (2) becomes

(3) $$\mathbb{A}0 = \mathbb{A} \cdot 10.$$

Numbers of the form (2) or (3) are called two-digit numbers (provided $\mathbb{A}$ is not 0).

The following table is organized so that each of its entries is the sum of the first number in the corresponding row and the top number in the corresponding column. This is the *addition table*.

+	1	2	3	4	5	6	7	8	9
1	2	3	4	5	6	7	8	9	10
2	3	4	5	6	7	8	9	10	11
3	4	5	6	7	8	9	10	11	12
4	5	6	7	8	9	10	11	12	13
5	6	7	8	9	10	11	12	13	14
6	7	8	9	10	11	12	13	14	15
7	8	9	10	11	12	13	14	15	16
8	9	10	11	12	13	14	15	16	17
9	10	11	12	13	14	15	16	17	18

Almost everyone has known the whole addition table since childhood, but very few people think of deducing its entries rigorously from axioms I–IV. Here is how it can be done.

The entries of the second column are justified by (1). Next note that if $\mathbb{A}$ is an arbitrary digit (different from 0), then

(4) $$10 + \mathbb{A} = 1\mathbb{A},$$

according to (2). For example, $10 + 7 = 17$.

We shall still need the formula

(5) $$(x + 1) + (y - 1) = x + y$$

for $x = 1, 2, 3, 4, 5, 6, 7, 8, 9$ and $y = 2, 3, 4, 5, 6, 7, 8, 9$, which follows from the fact that both sides are equal to $x + (1 + (y - 1))$, by the associativity of addition and the definition of difference.

The following calculations are based on formulas (1), (2), and (5):

$$2 + 2 = 3 + 1 = 4,$$
$$3 + 2 = 4 + 1 = 5,$$
$$3 + 3 = 4 + 2 = 5 + 1 = 6,$$
$$4 + 3 = 5 + 2 = 6 + 1 = 7,$$
$$4 + 4 = 5 + 3 = 6 + 2 = 7 + 1 = 8,$$

$$5 + 4 = 6 + 3 = 7 + 2 = 8 + 1 = 9,$$
$$5 + 5 = 6 + 4 = 7 + 3 = 8 + 2 = 9 + 1 = 10,$$
$$6 + 5 = 7 + 4 = 8 + 3 = 9 + 2 = 10 + 1 = 11,$$
$$6 + 6 = 7 + 5 = 8 + 4 = 9 + 3 = 10 + 2 = 12,$$
$$7 + 6 = 8 + 5 = 9 + 4 = 10 + 3 = 13,$$
$$7 + 7 = 8 + 6 = 9 + 5 = 10 + 4 = 14,$$
$$8 + 7 = 9 + 6 = 10 + 5 = 15,$$
$$8 + 8 = 9 + 7 = 10 + 6 = 16,$$
$$9 + 8 = 10 + 7 = 17,$$
$$9 + 9 = 10 + 8 = 18.$$

Using these equalities, we can fill in all the entries of the addition table on its main diagonal and below. In view of the commutativity of addition, the remaining entries can be filled in, too. This completes the justification of the entire addition table.

The following table is organized similarly, but instead of sums there are products of numbers appearing at the beginning of the corresponding column and row. This is the *multiplication table*.

·	1	2	3	4	5	6	7	8	9
1	1	2	3	4	5	6	7	8	9
2	2	4	6	8	10	12	14	16	18
3	3	6	9	12	15	18	21	24	27
4	4	8	12	16	20	24	28	32	36
5	5	10	15	20	25	30	35	40	45
6	6	12	18	24	30	36	42	48	54
7	7	14	21	28	35	42	49	56	63
8	8	16	24	32	40	48	56	64	72
9	9	18	27	36	45	54	63	72	81

The second column is justified by (·). Note now that

$$x \cdot 2 = x(1 + 1) = x + x;$$

hence the third column of the multiplication table is identical with the main diagonal in the addition table. In order to find the remaining entries, we introduce two auxiliary theorems.

1.7.1 *Each number in the multiplication table (except the first two columns) is the sum of the next number on the left and the number at the beginning of the same row.*

1.7.2   *If $\mathbb{A}$, $\mathbb{B}$, $\mathbb{C}$, and $\mathbb{D}$ denote digits, $\mathbb{D}$ being different from 0 and 9, then*

$$\mathbb{DA} + \mathbb{B} = \begin{cases} \mathbb{DC} & \text{if } \mathbb{A} + \mathbb{B} = \mathbb{C} \\ \mathbb{EC} & \text{if } \mathbb{A} + \mathbb{B} = 1\mathbb{C} \end{cases}$$

where $\mathbb{E} = \mathbb{D} + 1$.

For instance: $56 + 3 = 59$, since $6 + 3 = 9$, and
$56 + 8 = 64$, since $6 + 8 = 14$.

Proof of 1.7.1.   Assume that a number x stands in a row whose first element is $\mathbb{A}$ and in the column with the top element $\mathbb{B}$ ($\mathbb{B} \neq 1$). Then $x = \mathbb{A} \cdot \mathbb{B}$. The number on the left next to x is $\mathbb{A} \cdot (\mathbb{B} - 1)$, because it is in the row with the initial number $\mathbb{A}$ and in the column with the top number $\mathbb{B} - 1$. We thus have to prove that

$$\mathbb{A} \cdot \mathbb{B} = \mathbb{A} \cdot (\mathbb{B} - 1) + \mathbb{A}.$$

In fact,

$$\mathbb{A} \cdot \mathbb{B} = \mathbb{A} \cdot \big((\mathbb{B} - 1) + 1\big) = \mathbb{A} \cdot (\mathbb{B} - 1) + \mathbb{A} \cdot 1 = \mathbb{A} \cdot (\mathbb{B} - 1) + \mathbb{A}.$$

Proof of 1.7.2.   If $\mathbb{A} + \mathbb{B} = \mathbb{C}$, then

$$\mathbb{DA} + \mathbb{B} = (\mathbb{D} \cdot 10 + \mathbb{A}) + \mathbb{B} = \mathbb{D} \cdot 10 + (\mathbb{A} + \mathbb{B}) = \mathbb{D} \cdot 10 + \mathbb{C} = \mathbb{DC}.$$

If $\mathbb{A} + \mathbb{B} = 1\mathbb{C}$ and $\mathbb{E} = \mathbb{D} + 1$, then

$$\begin{aligned}
\mathbb{DA} + \mathbb{B} &= (\mathbb{D} \cdot 10 + \mathbb{A}) + \mathbb{B} = \mathbb{D} \cdot 10 + (\mathbb{A} + \mathbb{B}) \\
&= \mathbb{D} \cdot 10 + 1\mathbb{C} = \mathbb{D} \cdot 10 + (1 \cdot 10 + \mathbb{C}) = (\mathbb{D} \cdot 10 + 1 \cdot 10) + \mathbb{C} \\
&= (\mathbb{D} + 1) \cdot 10 + \mathbb{C} = \mathbb{E} \cdot 10 + \mathbb{C} = \mathbb{EC}.
\end{aligned}$$

Applying theorems 1.7.1 and 1.7.2 and using the addition table and the entries already established in the multiplication table, we successively find

$$\begin{array}{ll}
3 \cdot 3 = 6 + 3 = 9, & 7 \cdot 7 = 42 + 7 = 49, \\
4 \cdot 3 = 8 + 4 = 12, & 8 \cdot 3 = 16 + 8 = 24, \\
4 \cdot 4 = 12 + 4 = 16, & 8 \cdot 4 = 24 + 8 = 32, \\
5 \cdot 3 = 10 + 5 = 15, & 8 \cdot 5 = 32 + 8 = 40, \\
5 \cdot 4 = 15 + 5 = 20, & 8 \cdot 6 = 40 + 8 = 48, \\
5 \cdot 5 = 20 + 5 = 25, & 8 \cdot 7 = 48 + 8 = 56, \\
6 \cdot 3 = 12 + 6 = 18, & 8 \cdot 8 = 56 + 8 = 64, \\
6 \cdot 4 = 18 + 6 = 24, & 9 \cdot 3 = 18 + 9 = 27, \\
6 \cdot 5 = 24 + 6 = 30, & 9 \cdot 4 = 27 + 9 = 36, \\
6 \cdot 6 = 30 + 6 = 36, & 9 \cdot 5 = 36 + 9 = 45, \\
7 \cdot 3 = 14 + 7 = 21, & 9 \cdot 6 = 45 + 9 = 54, \\
7 \cdot 4 = 21 + 7 = 28, & 9 \cdot 7 = 54 + 9 = 63, \\
7 \cdot 5 = 28 + 7 = 35, & 9 \cdot 8 = 63 + 9 = 72, \\
7 \cdot 6 = 35 + 7 = 42, & 9 \cdot 9 = 72 + 9 = 81.
\end{array}$$

In this way all the entries on the diagonal and below have been found. By the commutativity of multiplication, the remaining part of the multiplication table may be filled up symmetrically with respect to the main diagonal.

Exercises 1.7

The notation described in this section is based on the number 10. There is no mathematical reason for using that particular number. Other integers can be used for the same purpose. In the following two problems you are asked to construct addition and multiplication tables in the notations based on 2 and 12.

1. We use only two digits: 0 and 1. We have $1 + 1 = 10$, and, in general,

$$\mathbb{A}\mathbb{B} = \mathbb{A} \cdot (1 + 1) + \mathbb{B},$$

$$\mathbb{A}\mathbb{B}\mathbb{C} = \mathbb{A} \cdot (1 + 1) \cdot (1 + 1) + \mathbb{B} \cdot (1 + 1) + \mathbb{C},$$

and so on. Construct the addition table and the multiplication table, including all three-digit binary numbers.

2. Construct the addition table and the multiplication table in the notation based on the number 12. Include all two-digit numbers.

CHAPTER 2

REAL NUMBERS

2.1 ADDITION

Natural numbers are a perfect tool for counting objects but generally they are insufficient for measuring quantities like distance, time, and weight. In order to describe phenomena involving such quantities, it is necessary to introduce real numbers.

We assume at first no link between real numbers and the natural numbers considered in the preceding chapter. The real numbers will be simply called *numbers*. The following are the fundamental properties of the *addition* of numbers:

1^+ $a + b = b + a$;

2^+ $(a + b) + c = a + (b + c)$;

3^+ The equation $a + x = b$ is solvable.

Equalities 1^+ and 2^+ hold for all numbers a, b, and c. Property 3^+ says that for every pair of numbers a and b there exists a number x satisfying $a + x = b$. We do not prove properties 1^+, 2^+, and 3^+, so they can be considered as axioms.

2.1.1 Theorem   *There exists one and only one number x such that, for all b,* $b + x = b$.

Proof.   The existence of a number x satisfying $b + x = b$ follows from 3^+. However, we may only assert that for every b there exists some number x satisfying the required equation, because we do not know whether that number is the same for all b. This must be proved.

Let a denote an arbitrary number. There exists a number x satisfying

$$(1) \qquad\qquad a + x = a.$$

We shall prove that the same number x satisfies $b + x = b$ for every number b. In fact, from 3^+ it follows that there is a number y such that $a + y = b$, or, equivalently,

$$(2) \qquad\qquad b = a + y.$$

Hence

$$b + x = (a + y) + x = a + (y + x) = a + (x + y) = (a + x) + y = a + y = b.$$

We thus have proved that the number x satisfying $a + x = a$ for some arbitrary a, satisfies the same equation for any a. We still have to prove that this is the only number with this property.

Assume that there are two such numbers and denote them by x_1 and x_2. Then we have simultaneously

$$x_1 + x_2 = x_1 \quad \text{and} \quad x_2 + x_1 = x_2.$$

From the first equality we have $x_1 = x_1 + x_2$ and from the second $x_2 = x_1 + x_2$, by 1^+. Hence $x_1 = x_2$. This proves that there is only one number with the required property.

Theorem 2.1.1 is interesting for the reason that it allows one to identify in the set of all real numbers exactly one real number with the following property: if we add that number to any number a, then as the result we obtain the same number a. The number with this property is called *zero* and is denoted by 0. In view of 1^+ we have

$$0 + a = a + 0 = a.$$

2.1.2 *If $a + x = b$ and $a + y = b$, then $x = y$.*

In other words: *The solution of $a + x = b$ is unique for each pair of numbers a and b.*

Proof.   If $a + x = b$ and $a + y = b$, then

(3) $a + x = a + y.$

By 3^+ there is a number z such that

(4) $a + z = 0.$

From (3) and (4) we find successively

$$(a + x) + z = (a + y) + z;$$
$$(x + a) + z = (y + a) + z;$$
$$x + (a + z) = y + (a + z);$$
$$x + 0 = y + 0;$$
$$x = y.$$

Note that 2.1.2 implies the following useful property, which is often called the *cancellation law for addition:*

2.1.3 *If $a + x = a + y$, then $x = y$.*

Exercises 2.1

1.   Does the set $\{0\}$ satisfy properties 1^+, 2^+, and 3^+?

2.   Denote $\mathbb{Z}_4 = \{0, 1, 2, 3\}$. Define addition in $\mathbb{Z}_4$ by the following table:

⊕	0	1	2	3
0	0	1	2	3
1	1	2	3	0
2	2	3	0	1
3	3	0	1	2

Does $\mathbb{Z}_4$ with addition $\oplus$ satisfy 1^+, 2^+, and 3^+?

3. Denote $\mathbb{Z}_5 = \{0, 1, 2, 3, 4\}$. Define addition in $\mathbb{Z}_5$ by the following table:

⊕	0	1	2	3	4
0	0	1	2	3	4
1	1	2	3	4	0
2	2	3	4	0	1
3	3	4	0	1	2
4	4	0	1	2	3

Does $\mathbb{Z}_5$ with addition $\oplus$ satisfy 1^+, 2^+, and 3^+?

2.2 SUBTRACTION

The unique solution x of the equation $a + x = b$ is called the *difference of a and b* and is denoted by $b - a$. From this definition it follows that

2.2.1 $a + (b - a) = b$.

By introducing the difference we have defined a new operation which associates a number $b - a$ with a pair of numbers a and b. This operation is called *subtraction*. Theorem 2.1.2 asserts the uniqueness of subtraction and property 3^+ its feasibility.

2.2.2 $a - a = 0$.

Proof.   By the definition of 0 we have

$$a + 0 = a$$

and by 2.2.1

$$a + (a - a) = a.$$

Hence $a - a = 0$, by 2.1.2.

2.2.3 $(b - a) + a = b$ *and* $(b + a) - a = b$.

Proof.   The first of these equalities follows at once from 2.2.1, by 1^+. In view of 2.2.1 we also may write $a + ((b + a) - a) = b + a$. Hence, by 1^+, we get $((b + a) - a) + a = b + a$ and finally $(b + a) - a = b$, by 2.1.3.

The number $0 - a$ is denoted simply by $-a$ and is called the *opposite of a*. We thus have

(1) $$0 - a = -a.$$

2.2.4 $a + (-a) = 0.$

Proof.   Substitute $b = 0$ in 2.2.1 and use (1).

2.2.5 $b + (-a) = b - a.$

Proof.   From 2.2.4 we have $(a + (-a)) + b = 0 + b = b$ and consequently $a + (b + (-a)) = b$. Comparing this with 2.2.1 we see that $b + (-a) = b - a$, by 2.1.2.

2.2.6 $-(-b) = b.$

Proof.   Substituting b for a in 2.2.4 we get $b + (-b) = 0$ and therefore, by 1^+, $(-b) + b = 0$. On the other hand, substituting $-b$ for a into the same equation 2.2.4 we obtain

$$(-b) + (-(-b)) = 0.$$

Consequently $-(-b) = b$, by 2.1.2.

Exercise 2.2

1.   Prove the following:

    (a) $a - 0 = a.$

    (b) $a + (b - c) = (a + b) - c.$

    (c) $a - (b - c) = (a + c) - b.$

    (d) $(a - b) + (c - d) = (a + c) - (b + d).$

    (e) $-(a + b) = -a - b.$

    (f) $-(b - a) = a - b.$

2.3 MULTIPLICATION

Multiplication assigns to each pair of numbers a and b another number ab, called the *product of a and b*. Multiplication has properties similar to those of summation:

    1^* $ab = ba$;

    2^* $(ab)c = a(bc)$;

    3^* The equation $ax = b$ is solvable whenever $a \neq 0$.

Property 3^* differs from the analogous property of addition by the supplementary assumption that a is different from 0. The necessity of this restriction is discussed in Section 2.5.

2.3.1 Theorem   *There exists one and only one number x such that, for all b,* $bx = b$.

Proof.   Let a denote an arbitrary number different from 0. In view of 3^* there exists a number x such that

(1) $$ax = a.$$

We are going to show that the same number x satisfies also $bx = b$ for all b.

By 3^* there exists a number y satisfying $ay = b$, or

(2) $$b = ay.$$

Hence

$$bx = (ay)x = a(yx) = a(xy) = (ax)y = ay = b.$$

We have proved existence of a number with the desired property. It remains to show its uniqueness. If there were two such numbers, for example, x_1 and x_2, then we would have

$$x_1 x_2 = x_1 \quad \text{and} \quad x_2 x_1 = x_2.$$

By 1^* the second equation can be written as $x_1 x_2 = x_2$; thus it follows that $x_1 = x_2$. This proves that the number is unique.

Theorem 2.3.1 distinguishes a number with the property that each number b multiplied by that number gives b as the result. This particular number is called the *unit* and is denoted by 1. In view of 1^* and 2.3.1 we have

$$1 \cdot a = a \cdot 1 = a.$$

2.3.2 *If $ax_1 = b$, $ax_2 = b$, and $a \neq 0$, then $x_1 = x_2$.*

In other words: *The solution of the equation $ax = b$ with $a \neq 0$ is unique.*

Proof.   If $ax_1 = b$ and $ax_2 = b$, then

(3) $$ax_1 = ax_2.$$

By 3^* there is a number y such that

(4) $$ay = 1.$$

From (3) and (4) we find successively

$$(ax_1)y = (ax_2)y,$$
$$(x_1 a)y = (x_2 a)y,$$
$$x_1(ay) = x_2(ay),$$
$$x_1 \cdot 1 = x_2 \cdot 1,$$
$$x_1 = x_2.$$

As in the case of addition, the above property implies the *cancellation law for multiplication:*

2.3.3 *If $ax = ay$ and $a \neq 0$, then $x = y$.*

Exercises 2.3

1. Does the set {0} satisfy 1*, 2*, and 3*?
2. Denote $\mathbb{Z}_4 = \{0, 1, 2, 3\}$. Define multiplication in $\mathbb{Z}_4$ by the following table:

$\otimes$	**0**	**1**	**2**	**3**
0	0	0	0	0
1	0	1	2	3
2	0	2	0	2
3	0	3	2	1

Does $\mathbb{Z}_4$ with multiplication $\otimes$ satisfy 1*, 2*, and 3*?

3. Denote $\mathbb{Z}_5 = \{0, 1, 2, 3, 4\}$. Define multiplication in $\mathbb{Z}_5$ by the following table:

$\otimes$	**0**	**1**	**2**	**3**	**4**
0	0	0	0	0	0
1	0	1	2	3	4
2	0	2	4	1	3
3	0	3	1	4	2
4	0	4	3	2	1

Does $\mathbb{Z}_5$ with multiplication $\otimes$ satisfy 1*, 2*, and 3*?

2.4 DIVISION

In view of 3* and 2.3.1 there exists, for each pair of numbers a and b such that $a \neq 0$, a number x satisfying the equation $ax = b$. This number is called the *quotient of the numbers b and a* or a *fraction with denominator a and numerator b*. This number is denoted by $\dfrac{b}{a}$ or b/a.

Since the number $\dfrac{b}{a}$ satisfies the equation $ax = b$, we have

(1) $$a\frac{b}{a} = b \quad \text{for} \quad a \neq 0.$$

Note that the above defines a new operation which assigns to every pair of numbers a and b such that $a \neq 0$ the number $\dfrac{b}{a}$. This operation is called *division*.

2.4.1 $\dfrac{a}{a} = 1$ *for* $a \neq 0$.

Proof. For $b = a$, (1) implies $a\dfrac{a}{a} = a$. On the other hand, $a \cdot 1 = a$. Hence, by 2.3.2, we obtain $\dfrac{a}{a} = 1$.

2.4.2 $c\dfrac{b}{a} = \dfrac{b}{a}c = \dfrac{bc}{a}$.

Proof. Multiplying both sides of (1) by c we get $\left[a \dfrac{b}{a} \right] c = bc$, and hence

$$(2) \qquad a \left(\frac{b}{a} c \right) = bc,$$

by 2*. But formula (1) holds for each number b, so it remains true when b is replaced by bc:

$$(3) \qquad a \frac{bc}{a} = bc.$$

From (2) and (3) it follows, by 2.3.2, that $\dfrac{b}{a} c = \dfrac{bc}{a}$.

2.4.3 $\dfrac{b}{a} a = b$ and $\dfrac{ba}{a} = b$ for $a \neq 0$.

Proof. The first equality follows from from (1) and 1*. If in (1) we write ba instead of b, then we get $a \dfrac{ba}{a} = ba$. On the other hand we have $ab = ba$. Hence the equation $\dfrac{ba}{a} = b$ follows by 2.3.2.

The number $\dfrac{1}{a}$, for $a \neq 0$, is called the *reciprocal* of a.

2.4.4 $c \dfrac{1}{a} = \dfrac{c}{a}$ for $a \neq 0$.

Proof. For $b = 1$ we infer from 2.4.2 that $c \dfrac{1}{a} = \dfrac{c \cdot 1}{a}$ and hence $c \dfrac{1}{a} = \dfrac{c}{a}$.

2.5 DISTRIBUTIVITY

We now complete the list of the properties of addition and multiplication with the following property which links the two operations:

$(+*)$ $a(b + c) = ab + ac$.

This property is called *distributivity of multiplication over addition.* In view of 1* we may also write

$$(b + c)a = ba + ca.$$

2.5.1 $a(b - c) = ab - ac$ and $(b - c)a = ba - ca$.

Proof. In view of 2.2.1 we have

$$(1) \qquad ac + (ab - ac) = ab;$$

$$(2) \qquad c + (b - c) = b.$$

Multiplying (2) by a we get

$$a(c + (b - c)) = ab$$

and applying $(+*)$

$$(3) \qquad ac + a(b - c) = ab.$$

From (1) and (3) we get $a(b - c) = ab - ac$, according to 2.1.2. In this way the first of equations 2.5.1 is proved. The second equation follows by 1^*.

2.5.2 $0 \cdot a = a \cdot 0 = 0$.

Proof. If $c = b$ in the first equation in 2.5.1 we obtain

$$a(b - b) = ab - ab$$

and hence $a \cdot 0 = 0$, by 2.2.2. That $0 \cdot a = 0$ follows by 1^*.

From 2.5.2 it follows that the equation $ax = b$ has no solution if $a = 0$ and $b \neq 0$. This explains the necessity of the restriction $a \neq 0$ in 3^*. On the other hand, if $a = 0$ and $b = 0$, then every number x satisfies $ax = b$. This explains the necessity of the restriction $a \neq 0$ in 2.3.2.

2.5.3 $(-a)b = -(ab)$ *and* $(-a)(-b) = ab$.

Proof. $(-a)b = (0 - a)b = 0 \cdot b - ab = 0 - ab = -ab$. Thus $(-a)b = -(ab)$ holds. Using this equality we may write

$$(-a)(-b) = -a(-b) = -a(0-b) = -(a \cdot 0 - ab) = -(0 - ab) = -(-ab) = ab,$$

by 2.5.1, 2.5.2, and 2.2.6.

2.5.4 *If* $a \neq 0$ *and* $b \neq 0$, *then* $ab \neq 0$.

Proof. It suffices to prove that if $a \neq 0$ and $ab = 0$, then $b = 0$. Indeed, since $a \cdot 0 = 0$ (by 2.5.2), the equality $b = 0$ follows from 2.3.2.

2.5.5 *If* $a \neq 0$ *and* $b \neq 0$, *then* $\dfrac{b}{a} \neq 0$.

Proof. It suffices to prove that if $a \neq 0$ and

(4)
$$\frac{b}{a} = 0,$$

then $b = 0$. Multiplying (4) by a we get $\dfrac{b}{a} a = 0 \cdot a$. Hence $b = 0$, by 2.4.3 and 2.5.2.

2.5.6 $\dfrac{b}{a} \cdot \dfrac{d}{c} = \dfrac{bd}{ac}$ *for* $a \neq 0$ *and* $c \neq 0$.

Proof. We have

$$(ac)\left(\frac{b}{a} \cdot \frac{d}{c}\right) = \left((ac)\frac{b}{a}\right)\frac{d}{c} = \left(a\left(c\frac{b}{a}\right)\right)\frac{d}{c} = \left(a\left(\frac{b}{a}c\right)\right)\frac{d}{c} = \left(\left(a\frac{b}{a}\right)c\right)\frac{d}{c}$$
$$= \left(a\frac{b}{a}\right)\left(c\frac{d}{c}\right);$$

hence

$$(ac)\left(\frac{b}{a} \cdot \frac{d}{c}\right) = bd.$$

On the other hand we may write

$$(ac)\frac{bd}{ac} = bd,$$

because ac is different from zero, by 2.5.4. From the last two equalities it follows that

$$\frac{b}{a} \cdot \frac{d}{c} = \frac{bd}{ac}.$$

2.5.7. $\dfrac{b}{a} = \dfrac{bc}{ac}$ *for $a \neq 0$ and $c \neq 0$.*

Proof.

$$\frac{b}{a} = \frac{b}{a} \cdot 1 = \frac{b}{a} \cdot \frac{c}{c} = \frac{bc}{ac},$$

by 2.4.1 and 2.5.6.

2.5.8. $\dfrac{b}{a} + \dfrac{c}{a} = \dfrac{b+c}{a}$ *and* $\dfrac{b}{a} - \dfrac{c}{a} = \dfrac{b-c}{a}$ *for $a \neq 0$.*

Proof. We have

$$\frac{b}{a} + \frac{c}{a} = b\frac{1}{a} + c\frac{1}{a} = (b+c)\frac{1}{a} = \frac{b+c}{a},$$

which proves the first formula. The proof of the second one is similar.

Formulas 2.5.7 and 2.5.8 allow us to add and subtract fractions with arbitrary denominators different from zero. In fact,

$$\frac{b}{a} + \frac{d}{c} = \frac{bc}{ac} + \frac{ad}{ac} = \frac{bc+ad}{ac} \quad \text{and} \quad \frac{b}{a} - \frac{d}{c} = \frac{bc}{ac} - \frac{ad}{ac} = \frac{bc-ad}{ac}.$$

2.5.9. $\dfrac{c}{\dfrac{b}{a}} = c\dfrac{a}{b}$ *for $a \neq 0$ and $b \neq 0$.*

Proof. By 2.5.4 we have $ab \neq 0$ and thus

$$1 = \frac{ab}{ab} = \frac{ba}{ab} = \frac{b}{a} \cdot \frac{a}{b}$$

and hence

$$\left(\frac{b}{a} \cdot \frac{a}{b}\right)c = 1 \cdot c = c$$

and

(1)
$$\frac{b}{a}\left(c\frac{a}{b}\right) = c.$$

Since $b/a \neq 0$, we may write

(2)
$$\frac{b}{a} \cdot \frac{c}{\dfrac{b}{a}} = c,$$

by (1) in Section 2.4. From (1) and (2), by using property 2.3.2, we obtain the required formula.

Exercises 2.5

1. Prove that

$$\frac{\left(\dfrac{d}{c}\right)}{\left(\dfrac{b}{a}\right)} = \frac{ad}{bc} \quad \text{for} \quad a \neq 0,\ b \neq 0,\ \text{and}\ c \neq 0.$$

2. Does $\mathbb{Z}_4$ with addition $\oplus$ and multiplication $\otimes$ satisfy $(+*)$? (See Exercises 2.1.2 and 2.3.2.)

3. Does $\mathbb{Z}_5$ with addition $\oplus$ and multiplication $\otimes$ satisfy $(+*)$? (See Exercises 2.1.3 and 2.3.3.)

2.6 INEQUALITIES

All the properties of real numbers considered so far can be called algebraic. They concern the algebraic operations of addition and multiplication. We now include the nonalgebraic property of being *positive*. To denote that a is positive we write $0 < a$. We have the following properties:

$1^<$ For every number a one and only one of the following relations holds:

$$a = 0, \quad 0 < a, \quad \text{or} \quad 0 < -a.$$

$2^<$ If $0 < a$ and $0 < b$, then $0 < a + b$ and $0 < ab$.

A number a such that $0 < -a$ is called negative. Property $1^<$ says that *every number is either zero, positive, or negative.* Property $2^<$ tells us that *the sum and the product of two positive numbers are positive.*

2.6.1 If $0 < a$ and $0 < b$, then $0 < \dfrac{b}{a}$.

Proof.   If we had $\dfrac{b}{a} = 0$, then we would get $b = \dfrac{b}{a} \cdot a = 0 \cdot a = 0$, which contradicts $1^<$. If we had $0 < -\dfrac{b}{a}$, it would mean that $0 < \left(-\dfrac{b}{a}\right) a = -b$, which again contradicts $1^<$. Only one possibility remains: $0 < \dfrac{b}{a}$.

2.6.2 If $a \neq 0$, then $0 < aa$.

Proof.   If $a \neq 0$, then $0 < a$ or $0 < -a$, by $1^<$. If $0 < a$, then $0 < aa$, by $2^<$. If $0 < -a$, then $0 < (-a)(-a) = aa$, by 2.5.3.

2.6.3 $0 < 1$.

Proof.   By 2.6.2 we have $0 < 1 \cdot 1 = 1$.

If $0 < b - a$, we say that *a is less than b* and write $a < b$. We can also say that *b is greater than a* and write $b > a$. By this convention the statement $0 < b$ can be read in two ways: "the number b is positive" or, because $0 < b - 0$, "the number b is greater than zero."

2.6.4 (Trichotomy)   *For each pair of numbers a and b one and only one of the following relations holds:*

$$a = b, \quad a < b, \quad \text{or} \quad b < a.$$

The proof follows directly from the definition of inequality and from $2^<$.

2.6.5 (Transitivity)   *If $a < b$ and $b < c$, then $a < c$.*

Proof.   The inequalities $a < b$ and $b < c$ mean $0 < b - a$ and $0 < c - b$. Therefore, in view of $2^<$, we get $0 < (b - a) + (c - b) = c - a$.

Instead of "$a < b$ and $b < c$" we often write "$a < b < c$."

2.6.6 *If $a < b$, then $a + c < b + c$.*

Proof.   The inequality $a < b$ means $0 < b - a$. Since

$$b - a = (b + c) - (a + c),$$

we obtain $0 < (b + c) - (a + c)$, which yields $a + c < b + c$.

2.6.7 *If $a < b$ and $c < d$, then $a + c < b + d$.*

The proof follows from $2^<$ and the definition of inequality.

2.6.8 *If $a < b$ and $0 < c$, then $ac < bc$.*

Proof.   If $a < b$, then $0 < b - a$ and $0 < (b - a)c$, by $2^<$. Hence $0 < bc - ac$, and $ac < bc$.

2.6.9 *If $a < b$, $c < d$ and $0 < b$, $0 < c$, then $ac < bd$.*

Proof.   From 2.6.8 it follows that $ac < bc$ and $bc < bd$. Hence $ac < bd$.

Exercises 2.6

1.   Prove 2.6.4.

2.   Prove 2.6.7.

3.   Let $a, b \in \mathbb{Z}_4$. We will write $a < b$ if the same inequality holds for the natural numbers a and b. Does $\mathbb{Z}_4$ with operations $\oplus$ and $\otimes$, and inequality $<$ satisfy $1^<$, $2^<$? (See Exercises 2.1.2 and 2.3.2.)

4.   Let $a, b \in \mathbb{Z}_5$. We will write $a < b$ if the same inequality holds for the natural numbers a and b. Does $\mathbb{Z}_5$ with operations $\oplus$ and $\otimes$, and inequality $<$ satisfy $1^<$, $2^<$? (See Exercises 2.1.3 and 2.3.3.)

5. Is it possible to define in $\mathbb{Z}_4$ a relation "<" so that $1^<$, $2^<$ would be satisfied? How about $\mathbb{Z}_5$?

6. Let X be a set with addition ($+$), multiplication ($\cdot$), and order ($<$) such that 1^+, 2^+, 3^+, 1^*, 2^*, 3^*, ($+*$), $1^<$, and $2^<$ are satisfied. Prove that if X has at least two elements then it has infinitely many elements.

7. Does the set $\{0\}$ satisfy all conditions 1^+, 2^+, 3^+, 1^*, 2^*, 3^*, ($+*$), $1^<$, and $2^<$?

8. Does the set $\mathbb{N}$ of all natural numbers satisfy all conditions 1^+, 2^+, 3^+, 1^*, 2^*, 3^*, ($+*$), $1^<$, and $2^<$?

9. Does the set $\mathbb{Q}$ of all rational numbers satisfy all conditions 1^+, 2^+, 3^+, 1^*, 2^*, 3^*, ($+*$), $1^<$, and $2^<$?

2.7 BOUNDED SETS

A set of numbers is said to be *bounded from below* by a number m, if all its elements are greater than or equal to m. The number m is then called a *lower bound* of that set. If m is a lower bound, then each number less than m is another lower bound of the same set. For instance the set of all positive numbers is bounded from below. All negative numbers as well as zero are its lower bounds. The number 0 is the greatest lower bound. It is also the greatest lower bound of the set of all non-negative numbers. In the case of the non-negative numbers the greatest lower bound belongs to the considered set, whereas in the case of the positive numbers it does not.

Generally a number is called the *greatest lower bound* of a given set P, if

$\mathbf{I_{GLB}}$ $m \le x$ for each $x \in P$;

$\mathbf{II_{GLB}}$ m is the greatest of numbers satisfying $\mathrm{I_{GLB}}$; that is, if $n \le x$ for each $x \in P$, then $m \ge n$.

Notice that a set can have only one greatest lower bound. Indeed, assuming that m_1 and m_2 are the greatest lower bounds of P, we have $m_1 \le m_2$ and $m_2 \le m_1$, so $m_1 = m_2$, which proves uniqueness.

The question arises whether every set bounded from below possesses a greatest lower bound. This does not follow from the properties of real numbers considered so far (see Exercise 2.6.9). If we want to use this property, it must be adopted as a new axiom.

D For every nonempty set bounded from below there exists a greatest lower bound.

This axiom is often called the *Dedekind Axiom*. Richard Dedekind (1831–1916), was the first to construct a model of the real numbers.

All together we get the following set of 10 *Axioms of the Real Numbers:*

1^+ $\quad$ $a + b = b + a$.

2^+ $\quad$ $(a + b) + c = a + (b + c)$.

3^+ $\quad$ The equation $a + x = b$ is solvable.

1^* $\quad$ $ab = ba$.

2^* $\quad$ $(ab)c = a(bc)$.

3^* $\quad$ The equation $ax = b$ is solvable, whenever $a \neq 0$.

$(+*)$ $\quad$ $a(b + c) = ab + ac$.

$1^<$ $\quad$ For every number a one and only one of the following relations holds:

$$a = 0, \quad 0 < a, \quad \text{or} \quad 0 < -a.$$

$2^<$ $\quad$ If $0 < a$ and $0 < b$, then $0 < a + b$ and $0 < ab$.

D $\quad$ For every nonempty set bounded from below there exists a greatest lower bound.

We will refer to this set of ten axioms as $\mathfrak{R}$.

—Do these axioms entirely characterize real numbers?

—People trained in formal logic would say no. For the set consisting of the single element 0 satisfies all the ten axioms. And yet it is not the set of all real numbers.

—And if we assume that the set contains at least two elements?

—Then everything is all right. The existence of two numbers implies, by means of adopted axioms, the existence of all real numbers. But we have to be very careful with the words "two elements" which have to be understood in the colloquial sense: "one element and another one". A careful reader could notice that we have used the existence of two different real numbers in the proofs of 2.3.1 and 2.6.3. In the first case we need an element different from 0. In the second case we use the fact that $1 \neq 0$, which follows from the existence of two different elements.

—So the existence of at least two elements is to be adopted as an additional eleventh axiom of real numbers.

—This is one possibility. Or else we may say that the set of real numbers is the set containing at least two elements and satisfying axioms $\mathfrak{R}$.

—I like this, because then the number of axioms remains 10, and it is very nice to have exactly as many axioms as fingers.

A set is said to be *bounded from above* if all its elements are less than or equal to a number m. By the *least upper bound* of a given set P we mean a number m such that:

$\mathbf{I_{LUB}}$ $\quad$ $x \leq m$ for each $x \in P$;

$\mathbf{II_{LUB}}$ $\quad$ m is the least of numbers satisfying $\mathrm{I_{LUB}}$; that is, if $x \leq n$ for each $x \in P$, then $m \leq n$.

Using axiom D one can easily prove the following property:

D' *Each nonempty set bounded from above has a least upper bound.*

To see that D implies D' and, conversely, that D' implies D, it suffices to consider the set of the negatives of the elements of P; that is, a number x belongs to P' if and only if $-x$ belongs to P. Thus, axiom D can be equivalently replaced by D'.

In order to avoid possible misunderstanding it should be strongly emphasized that if we say "the set P has greatest lower bound m" (or "least upper bound m"), then m need not belong to P. It only means that a number with the required properties exists.

2.7.1 *For each pair of real numbers a and b one and only one of the following relations holds:*

$$a \le b \quad \text{or} \quad b < a.$$

Proof.   This is a consequence of 2.6.4.

2.7.2 If $a \le b \le a$, then $a = b$.

Proof.   The inequality $a \le b$ means that either $a = b$ or $a < b$; the inequality $b \le a$ means that $b = a$ or $b < a$. Since by 2.6.4 the inequalities $a < b$ and $b < a$ exclude each other, only one possibility remains: $a = b$.

Exercises 2.7

1. Provide the details of the proof of the equivalence of D and D'.
2. Find greatest lower bounds and least upper bounds of the following sets:
    (a) $\{1, \frac{1}{2}, \frac{1}{3}, \frac{1}{4}, \ldots\}$.
    (b) $\{0, -\frac{1}{2}, \frac{2}{3}, -\frac{3}{4}, \frac{4}{5}, -\frac{5}{6}, \ldots\}$.
3. Find the least upper bound of the set of all rational numbers whose square is less than 2.
4. Consider the set $X = \{a_1, a_2, a_3, \ldots\}$ where the a_n are defined inductively:

    $$a_1 = 1; \qquad a_{n+1} = \frac{3a_n + 4}{2a_n + 3} \quad \text{for} \quad n = 1, 2, 3, \ldots.$$

    Show that X is bounded from above and from below. Can you find the least upper bound of X?

2.8 CONNECTION BETWEEN REAL AND NATURAL NUMBERS

Assuming that there exists a set $\mathbb{R}$ with operations $+$ and $\cdot$, and inequality $<$, that satisfies axioms $\mathscr{R}$, we can select from the set $\mathbb{R}$ a nonempty subset $\mathbb{N}$ satisfying axioms I–IV formulated in Chapter 1. In fact, let us consider a subset with the following properties:

(1)   The number 1 belongs to the set;

(2)   If p belongs to the set, then so does $p + 1$.

These properties however do not characterize uniquely the set $\mathbb{N}$ of natural numbers, for there exist other sets satisfying (1) and (2), for instance the set of all real numbers. The set $\mathbb{N}$ may be distinguished by assuming that $\mathbb{N}$ contains only the number 1 and those numbers that can be obtained by successive addition of 1. $\mathbb{N}$ is therefore the smallest set satisfying (1) and (2). To be precise, we say that $\mathbb{N}$ contains each number that belongs to every set Z having properties (1) and (2). The set $\mathbb{N}$ so defined continues to have properties (1) and (2). Indeed, the number 1 belongs to $\mathbb{N}$, since it belongs to every Z. Moreover, if p belongs to $\mathbb{N}$, then p belongs to every Z. Thus $p + 1$ belongs to every Z, and consequently $p + 1$ belongs to $\mathbb{N}$.

Now, since the set $\mathbb{N}$ is defined as a subset of the real numbers, the inequality relation is already defined for $\mathbb{N}$. It can be proved that the set $\mathbb{N}$ and inequality $<$ satisfy axioms I–IV. We shall not discuss the details for two reasons. First, the proofs are boring. Second, it does not seem to be important to deduce the simple and almost obvious statements I–IV from a far more complicated system of axioms $\mathscr{R}$. It would be much more interesting to deduce, from the simple axioms I–IV, the existence of a set $\mathbb{R}$ satisfying axioms $\mathscr{R}$. It is amazing that this is possible.

A set $\mathbb{R}$ with operations $+$ and $\cdot$, and inequality $<$ satisfying axioms $\mathscr{R}$, is called a model of real numbers. Note that in $\mathscr{R}$ there are three existential axioms: 3^+, 3^*, and D. Axiom D is the only one of these three that is satisfied by the set of natural numbers $\mathbb{N}$. In order to obtain a model for the real numbers we have to extend the concept of numbers in three steps. First we introduce integers so that 3^+ is satisfied. Then we introduce rational numbers so that 3^+ and 3^* hold. There is still a third step needed, because property D is lost when the rational numbers are introduced.

The ancients knew integers and rational numbers well, but they had difficulties with irrational numbers. For instance, in the Pythagorean school they did not know of any number that could be assigned to the length of the diagonal of a unit square. It was forbidden to speak to other people about it. Not until the nineteenth century was a model for real numbers constructed, by Dedekind. Other models were constructed later. In this book we shall use the method of Georg Cantor (1845–1918). It is based on the concept of equivalence classes, which is also needed when correctly introducing integers and rational numbers.

2.9 EQUIVALENCE CLASSES

The method of equivalence classes enables us to construct new mathematical objects from given ones. It is a very powerful tool in mathematics and we shall discuss it in its full generality.

2.9.1 Definition   A relation $x \sim y$ defined on a set X is called an *equivalence* if it has the following properties:

(a) $x \sim y$ implies $y \sim x$ (*symmetry*);

(b) $x \sim y$ and $y \sim z$ implies $x \sim z$ (*transitivity*).

If $x \sim y$, then we say x *is equivalent to* y. In view of (a), we may also say y is equivalent to x. The set of all elements in X that are equivalent to an element $x \in X$ is called the *equivalence class* of x and is denoted by $[x]$. It may happen that an element x is not equivalent to any element in X, not even itself. Then $[x]$ is empty. On the other hand, if $[x]$ is not empty, then $x \sim x$ and therefore $x \in [x]$. In fact, if $y \in [x]$, then $x \sim y$ and $y \sim x$, and hence $x \sim x$. An element of an equivalence class is called a *representative* of that class.

2.9.2 Theorem

(1) *If $y \sim z$, then $[y] = [z]$; that is, the classes $[y]$ and $[z]$ have the same elements.*

(2) *If the relation $y \sim z$ does not hold, then the classes $[y]$ and $[z]$ are disjoint; that is, have no common elements.*

Proof.   To prove (1) suppose $y \sim z$. If $x \in [y]$, then $x \sim y$. Thus $x \sim z$, in view of (b), which implies $x \in [z]$. Therefore $[y] \subset [z]$. Similarly we prove that $[z] \subset [y]$.

To prove (2) suppose that the relation $y \sim z$ does not hold and that there exists an element x belonging to both $[y]$ and $[z]$. Then $x \sim y$ and $x \sim z$ and, by (a) and (b), $y \sim z$, contrary to the hypothesis.

Forming equivalence classes is like sticking some elements together into bigger grains. Those elements not equivalent to any element drop out in the construction.

Exercises 2.9

1. Give an example of an equivalence relation for which some equivalence classes are empty.

2. Let $<$ be a relation on $\mathbb{N}$ defined as follows:

    $$x < y \quad \text{if there exists } z \in \mathbb{N} \text{ such that } xz = y.$$

    Is $<$ an equivalence relation?

3. Let $\triangle$ be a relation defined on $\mathbb{N}$ as follows:

    $$x \triangle y \quad \text{if there exist } m, n, k \in \mathbb{N}, \ k \neq 1, \quad \text{such that } x = mk \text{ and } y = nk.$$

    Is $\triangle$ an equivalence relation?

4. Let n be a fixed natural number. We define the relation $=_n$ on $\mathbb{N}$ by

    $$a =_n b \quad \text{if } n \text{ divides } b - a.$$

    In other words, $a =_n b$ if there exists an integer k such that $b - a = kn$. Is the defined relation an equivalence? Characterize the equivalence classes.

5. Let X be the set of all pairs of natural numbers. Define the relation $\square$ on X as follows:

    $$(s, t) \square (u, v) \quad \text{if } s + t = u + v.$$

Is $\square$ an equivalence relation?

6. True or false? $y \sim z$ if and only if $[y] = [z]$.

2.10 DEFINING INTEGERS BY NATURAL NUMBERS

If $n < m$ where m and n are natural numbers, then $m - n$ is a natural number. However it is not so if $m \leq n$. In this case the difference is nothing but an ordered pair of natural numbers. If we considered different pairs as different numbers, then for instance the pairs $1 - 2$, $5 - 6$, $8 - 9$ would represent different numbers in contrast to what we have learned at school. The identification of these pairs can be correctly carried out by using the method of equivalence classes.

We say that two pairs $m_1 - n_1$ and $m_2 - n_2$ are equivalent if and only if $m_1 + n_2 = m_2 + n_1$; we then write $m_1 - n_1 \sim m_2 - n_2$. One can easily see that this relation is symmetric. To check that it is also transitive, assume that

$$m_1 - n_1 \sim m_2 - n_2 \quad \text{and} \quad m_2 - n_2 \sim m_3 - n_3.$$

Then

$$m_1 + n_2 = m_2 + n_1 \quad \text{and} \quad m_2 + n_3 = m_3 + n_2.$$

Summing these two equalities we obtain

$$m_1 + n_3 + (m_2 + n_2) = m_3 + n_1 + (m_2 + n_2),$$

and in view of the cancellation law,

$$m_1 + n_3 = m_3 + n_1.$$

This proves that $m_1 - n_1 \sim m_3 - n_3$.

The equivalence classes of pairs of natural numbers are a model of the set of integers. This model contains all natural numbers if we agree with the following identification:

$$n = [(n + 1) - 1].$$

Now we introduce for integers addition, multiplication, and the property of being positive.

$$[m_1 - n_1] + [m_2 - n_2] = [(m_1 + m_2) - (n_1 + n_2)];$$

$$[m_1 - n_1][m_2 - n_2] = [(m_1 m_2 + n_1 n_2) - (m_1 n_2 + m_2 n_1)];$$

$$[m - n] > 0 \text{ if and only if } m > n.$$

In the inequality $[m - n] > 0$, 0 is the equivalence class $[n - n]$.

Of course, we need proofs showing that these operations are well defined and that the usual properties of addition, multiplication, and inequalities are satisfied. This matter will be discussed together with rational numbers. See also the following exercises.

Exercises 2.10

In the following exercises, be careful to use only the properties of natural numbers.

1. Let $(m_1 - n_1) \sim (k_1 - l_1)$ and $(m_2 - n_2) \sim (k_2 - l_2)$. Prove that
$$(m_1 + m_2) - (n_1 + n_2) \sim (k_1 + k_2) - (l_1 + l_2).$$

2. Let $(m_1 - n_1) \sim (k_1 - l_1)$ and $(m_2 - n_2) \sim (k_2 - l_2)$. Prove that
$$(m_1 m_2 + n_1 n_2) - (m_1 n_2 + m_2 n_1) \sim (k_1 k_2 + l_1 l_2) - (k_1 l_2 + k_2 l_1).$$

3. Let $(m - n) \sim (k - l)$. Prove that
$$m > n \text{ if and only if } k > l.$$

4. Prove that for any pair of integers a and b the equation $a + x = b$ has a solution.

5. Prove that for any integers a, b, and c we have $a(b + c) = ab + ac$.

6. Prove that, for any pair of integers a and b, $a > 0$ and $b > 0$ together imply that $a + b > 0$ and $ab > 0$.

2.11 DEFINING RATIONAL NUMBERS BY INTEGERS

The construction of rational numbers is similar to the construction of integers, but the relation of equivalence is different. We consider the set of all pairs (p, q), where p and q are integers and $q \neq 0$. Since those pairs correspond to fractions we write p/q instead of (p, q).

Two pairs p/q and r/s are considered *equivalent* if $ps = rq$. We then write $p/q \sim r/s$. It is easy to check that this relation is symmetric and transitive, so it is an equivalence. The set of equivalence classes of this relation is a model of the set of rational numbers. The equivalence class containing a pair p/q is denoted by $[p/q]$.

We introduce the following definitions of addition, multiplication, and the property of being positive, respectively:

$$\left[\frac{p}{q} \right] + \left[\frac{r}{s} \right] = \left[\frac{ps + rq}{qs} \right];$$

$$\left[\frac{p}{q} \right] \left[\frac{r}{s} \right] = \left[\frac{pr}{qs} \right];$$

$$\left[\frac{p}{q} \right] > 0 \text{ if } pq > 0.$$

The set of rational numbers contains all integers if we agree with the identification $p = [p/1]$.

Now we have a model of rational numbers. This model satisfies all properties $\mathcal{R}$ except for D. Let us consider, for example, property 2^+. We first prove this property for integers. We have

$$\big([p - q] + [r - s]\big) + [t - u] = [(p + r) - (q + s)] + [t - u]$$
$$= [((p + r) + t) - ((q + s) + u))]$$

and

$$[p - q] + \big([r - s] + [t - u]\big) = [p - q] + [(r + t) - (s + u)]$$
$$= [(p + (r + t)) - (q + (s + u))].$$

In both cases the result can be written as

$$[(p + r + t) - (q + s + u)].$$

This proves 2^+ for integers. The brackets have been used here for equivalence classes in the sense of the preceding section. Using our result we shall now prove 2^+ for arbitrary rational numbers. In the following argument brackets will be used in the new sense as equivalence classes of fractions.

We have

$$\left(\left[\frac{p}{q}\right] + \left[\frac{r}{s}\right]\right) + \left[\frac{t}{u}\right] = \left[\frac{ps + rq}{qs}\right] + \left[\frac{t}{u}\right] = \left[\frac{(ps + rq)u + tqs}{qsu}\right]$$

and

$$\left[\frac{p}{q}\right] + \left(\left[\frac{r}{s}\right] + \left[\frac{t}{u}\right]\right) = \left[\frac{p}{q}\right] + \left[\frac{ru + ts}{su}\right] = \left[\frac{psu + (ru + ts)q}{qsu}\right].$$

In both cases the result can be written as

$$\left[\frac{psu + rqu + tqs}{qsu}\right]$$

so that 2^+ holds.

—This proof is very long, not interesting, and actually boring.

—And yet it is not complete.

—Why not?

—Because in the calculation with rational numbers we have used more properties of integers. In fact, first of all we should prove that the definitions of addition and multiplication of integers are correct. That is,

if $[m_1 - n_1] = [p_1 - q_1]$ and $[m_2 - n_2] = [p_2 - q_2]$,   then

$$[(m_1 + m_2) - (n_1 + n_2)] = [(p_1 + p_2) - (q_1 + q_2)]$$ and

$$[(m_1 m_2 + n_1 n_2) - (m_1 n_2 + m_2 n_1)] = [(p_1 p_2 + q_1 q_2) - (p_1 q_2 + p_2 q_1)].$$

Next, we have to prove properties 2^+, 1^*, 2^*, and $(+*)$ for integers, because they have also been used in the proof presented above. Finally we need to check that the definition of addition of rational numbers is correct. Only then would the proof be complete.

—So we first should prove all these properties for integers! There should not be any serious problem with that. Anyhow the proofs would be awfully boring and the complete proof of 2^+ for rational numbers would be twice as long, at least.

—This is right. The proofs of the first nine properties in $\mathscr{R}$ are like a big puzzle for children. One needs nothing but patience.

—Moreover, as those properties have been familiar since the elementary school, we strongly believe that they must be true.

—In our case we do not need belief, because we have shown a way which enables us to check the whole theory scientifically.

—So, should every genuine mathematician check all the items in $\mathscr{R}$?

—No, but every genuine mathematician should feel able to provide a proof for any property in $\mathscr{R}$. I think that no mathematician in the world has checked all the proofs, except perhaps those who were writing a book on arithmetic.

Exercises 2.11

In the following exercises, be careful to use only the properties of integers.

1. Let $p_1/q_1 \sim r_1/s_1$ and $p_2/q_2 \sim r_2/s_2$. Prove that

$$\frac{p_1 q_2 + p_2 q_1}{q_1 q_2} \sim \frac{r_1 s_2 + r_2 s_1}{s_1 s_2}.$$

2. Let $p_1/q_1 \sim r_1/s_1$ and $p_2/q_2 \sim r_2/s_2$. Prove that

$$\frac{p_1 p_2}{q_1 q_2} \sim \frac{r_1 r_2}{s_1 s_2}.$$

3. Let $p/q \sim r/s$. Prove that $pq > 0$ if and only if $rs > 0$.

2.12 DEFINING REAL NUMBERS BY RATIONAL NUMBERS

Real numbers, like integers and rational numbers, will be introduced by the method of equivalence classes. However it will not be sufficient to consider pairs. Instead we shall consider infinite sequences of rational numbers.

An example of a sequence is

$$(1) \qquad \frac{1}{1}, \ \frac{1}{2}, \ \frac{1}{3}, \ldots .$$

One can readily guess that the numbers that would follow are $\frac{1}{4}$, $\frac{1}{5}$, $\frac{1}{6}$, $\frac{1}{7}$, and so on. However, there is no mathematical or logical necessity for that. For instance, if we consider expression $\frac{1}{6}(n^2 - 6n + 11)$, then we obtain numbers in (1) for $n = 1$, 2, 3, but then we get $\frac{1}{2}$, 1, $\frac{11}{6}$, and 3 for $n = 4$, 5, 6, and 7. In order to describe sequence (1) in a way that does not raise any doubts we write

$$\left\{ \frac{1}{n} \right\} \quad (n \in \mathbb{N}).$$

Generally one can write

$$\{a_n\} \quad (n \in \mathbb{N})$$

which exhibits the fact that to each natural number n there corresponds an element a_n. Writing "$(n \in \mathbb{N})$" after every sequence would be rather cumbersome. For reasons of simplicity we will write $\{a_n\}$ or $\{1/n\}$ instead of $\{a_n\}$ $(n \in \mathbb{N})$ or $\{1/n\}$ $(n \in \mathbb{N})$.

2.12.1 Definition   Two sequences of rational numbers $\{a_n\}$ and $\{b_n\}$ are equivalent, denoted $\{a_n\} \sim \{b_n\}$, if for any rational number $\varepsilon > 0$ there exists a natural number K such that

$$-\varepsilon < a_m - b_n < \varepsilon \quad \text{for all } m, n > K.$$

The defined relation is obviously symmetric. To see that it is also transitive, assume that $\{a_n\} \sim \{b_n\}$ and $\{b_n\} \sim \{c_n\}$. For any $\varepsilon > 0$, there are numbers K_1 and K_2 such that

$$-\frac{\varepsilon}{2} < a_m - b_n < \frac{\varepsilon}{2} \quad \text{for all } m, n > K_1,$$

$$-\frac{\varepsilon}{2} < b_n - c_k < \frac{\varepsilon}{2} \quad \text{for all } n, k > K_2.$$

Adding these inequalities we obtain

$$-\varepsilon < a_m - c_k < \varepsilon \quad \text{for all } m, k > K,$$

where K is the greater of K_1 and K_2. This proves that $\sim$ is an equivalence.

For example, the sequences $\{1/n\}$ and $\{0\}$ are equivalent where $\{0\}$ denotes the sequence $\{a_n\}$ for which $a_n = 0$ for all $n \in \mathbb{N}$. The sequence $\{n\}$ is not equivalent to any sequence, not even to itself.

The equivalence class containing a sequence $\{a_n\}$ will be denoted by $[\{a_n\}]$. According to this definition the symbol $[\{n\}]$ represents the empty set, because there is no sequence of rational numbers equivalent to the sequence $\{n\}$. We will use a convention that the symbol $[\{a_n\}]$ is used only to represent a nonempty class. The set of all nonempty equivalence classes is a model of real numbers. Therefore, in this model, a real number is an equivalence class of sequences of rational numbers.

Note that if $[\{a_n\}]$ represents a real number, then $\{a_n\}$ is equivalent to itself (see the remarks in Section 2.9). Conversely, if $\{a_n\} \sim \{a_n\}$, then $[\{a_n\}]$ represents a real number, because $\{a_n\}$ belongs to $[\{a_n\}]$.

Clearly, for any rational number a the constant sequence $\{a\}$ belongs to an equivalence class: it is equivalent to itself and, for example, to $\{a + 1/n\}$. This equivalence class is denoted by $[\{a\}]$. We admit the identification $a = [\{a\}]$, so that each rational number is a real number.

Having so introduced real numbers, we have to define operations on them, and then prove all the properties $\mathcal{R}$. We adopt the following definitions of addition and multiplication:

$$[\{a_n\}] + [\{b_n\}] = [\{a_n + b_n\}] \quad \text{and} \quad [\{a_n\}][\{b_n\}] = [\{a_n b_n\}].$$

Note that there is a question here of well-definedness (see Exercises 2.12.6 and 2.12.7).

We have

$$([\{a_n\}] + [\{b_n\}]) + [\{c_n\}] = [\{a_n + b_n\}] + [\{c_n\}] = [\{(a_n + b_n) + c_n\}],$$

$$[\{a_n\}] + ([\{b_n\}] + [\{c_n\}]) = [\{a_n\}] + [\{b_n + c_n\}] = [\{a_n + (b_n + c_n)\}],$$

and in both cases the final result can be written as $[\{a_n + b_n + c_n\}]$. This proves property 1^+. We see that the proof is a little simpler than the one for rational numbers (of course we use here former results for rational numbers). The detailed verification of all the axioms $\mathscr{R}$ would be an idle task for the same reason as for rational numbers. The only interesting question is about axiom D.

The definition of inequalities for real numbers is somewhat more sophisticated:

2.12.2 Definition   We write $[\{a_n\}] > 0$, and say that $[\{a_n\}]$ is *positive,* if there exist a positive rational number ε and a natural number K such that $a_n > \varepsilon$ for all $n > K$. Similarly, we write $[\{a_n\}] < 0$, and say that $[\{a_n\}]$ is *negative,* if there exist a positive rational number ε and a natural number K such that $a_n < -\varepsilon$ for all $n > K$.

Again, there is a question here of well-definedness (see Exercise 2.12.8). Clearly, the inequality $[\{a_n\}] < 0$ is equivalent to $[\{-a_n\}] > 0$.

We shall prove property $1^<$. First note that the relations $a = 0$, $0 < a$, and $0 < -a$ exclude one another. Now let $[\{a_n\}]$ be a real number. Since $\{a_n\} \sim \{a_n\}$, for every $\varepsilon > 0$ there exists a number K such that

$$(2) \qquad -\frac{\varepsilon}{2} < a_m - a_n < \frac{\varepsilon}{2} \quad \text{for all } m, n > K.$$

If $[\{a_n\}]$ is not positive, then there exists an index $n > K$ such that $a_n < \frac{\varepsilon}{2}$. Hence, in view of (2) we get $a_m < \varepsilon$ for every $m > K$. Similarly, assuming that $[\{a_n\}]$ is not negative we get $a_m > -\varepsilon$ for every $m > K$. Thus, if $[\{a_n\}]$ is neither positive nor negative, we have

$$-\varepsilon < a_n < \varepsilon \quad \text{for all } n > K.$$

Since ε can be chosen arbitrarily small, this proves that $\{a_n\} \sim \{0\}$ and, consequently, $[\{a_n\}] = [\{0\}]$. Property $1^<$ is proved.

To prove $2^<$ assume that $[\{a_n\}] > 0$ and $[\{b_n\}] > 0$. There exist rational numbers $\varepsilon_1 > 0$ and $\varepsilon_2 > 0$ and numbers n_1 and n_2 such that

$$a_n > \varepsilon_1 \quad \text{for } n > n_1 \quad \text{and} \quad b_n > \varepsilon_2 \quad \text{for } n > n_2.$$

Then

$$a_n + b_n > \varepsilon_1 + \varepsilon_2 \quad \text{and} \quad a_n b_n > \varepsilon_1 \varepsilon_2 \quad \text{for} \quad n > n_0,$$

where n_0 is the greater of the numbers n_1, n_2. Since $\varepsilon_1 + \varepsilon_2 > 0$ and $\varepsilon_1 \varepsilon_2 > 0$, property $2^<$ is proved.

2.12.3 Definition   By $a < b$ and $a \leq b$, where a and b are real numbers, we mean that $b - a > 0$ and $b - a \geq 0$, respectively.

2.12.4 Theorem   *For every real number $\alpha > 0$ there is a rational number ε such that $0 < \varepsilon < \alpha$.*

Proof.   If $\alpha = [\{a_n\}] > 0$, then there exists a rational number $\varepsilon > 0$ and a natural number n_0 such that $a_n > 2\varepsilon$ for $n > n_0$. Hence $a_n - \varepsilon > \varepsilon$ for $n > n_0$, and thus $[\{a_n - \varepsilon\}] > 0$. But since ε is a rational number, we have $[\{a_n\}] - \varepsilon = [\{a_n - \varepsilon\}] > 0$; that is, $\alpha = [\{a_n\}] > \varepsilon$.

2.12.5 Theorem   *If* $a_1, a_2, \ldots$ *and* b *are rational numbers such that*

(3) $$a_n \le a_{n+1} < b \quad \text{for all} \quad n \in \mathbb{N},$$

then the sequence $\{a_n\}$ *represents a real number and* $[\{a_m, a_m, a_m, \ldots\}] \le [\{a_n\}]$ *for each* $m \in \mathbb{N}$. *In other words, every increasing sequence of rational numbers bounded from above represents a real number which is greater than or equal to every term of the sequence.*

Proof.   To prove that the sequence $\{a_n\}$ represents a real number it suffices to show that $\{a_n\} \sim \{a_n\}$, that is, for every rational number $\varepsilon > 0$ there exists a number n_0 such that

$$-\varepsilon < a_m - a_n < \varepsilon \quad \text{for } m, n > n_0.$$

Suppose this is not true. Then there exists a number $\varepsilon > 0$ such that

$$a_{p_n} - a_{q_n} \ge \varepsilon$$

for all $n \in \mathbb{N}$, with $p_n \to \infty$, $q_n \to \infty$. We may assume that $q_n < p_n < q_{n+1}$ for all $n \in \mathbb{N}$. Then

$$a_{p_n} - a_{q_1} = (a_{p_n} - a_{q_n}) + (a_{q_n} - a_{p_{n-1}}) + (a_{p_{n-1}} - a_{q_{n-1}}) + \ldots$$
$$+ (a_{q_2} - a_{p_1}) + (a_{p_1} - a_{q_1})$$
$$\ge (a_{p_n} - a_{q_n}) + (a_{p_{n-1}} - a_{q_{n-1}}) + \ldots$$
$$+ (a_{p_2} - a_{q_2}) + (a_{p_1} - a_{q_1}) \ge n\varepsilon,$$

because in the sum each of the $2n - 1$ differences is non-negative ($\{a_n\}$ is nondecreasing) and n of them are greater than ε. Thus $a_{p_n} \ge a_{q_1} + n\varepsilon$ for all $n \in \mathbb{N}$, which contradicts the assumption that the sequence is bounded above.

In order to prove $[\{a_m, a_m, a_m, \ldots\}] \le [\{a_n\}]$, suppose that, conversely, $[\{a_m, a_m, a_m, \ldots\}] > [\{a_n\}]$ for some $m \in \mathbb{N}$. Then $[\{a_m - a_n\}] > 0$; that is, $a_m > a_n$ for every n greater than some n_0, which contradicts the assumption that the sequence is increasing.

Now we are going to prove that our model of the real numbers satisfies D. Let P be a set of real numbers, bounded from below. We show that P has a greatest lower bound. Let q_1 be a fixed natural number and let p_1 be the greatest integer such that $p_1/q_1 \le x$ for all $x \in P$. If p_1/q_1 is the greatest lower bound of P, we have nothing more to do. If it is not, then we put $q_2 = 2q_1$ and denote by p_2 the greatest integer such that $p_2/q_2 \le x$ for all $x \in P$. Clearly $p_2/q_2 \ge p_1/q_1$. If p_2/q_2 is the greatest lower bound of P, then we have nothing more to do. If it is not, put $q_3 = 2q_2$ and denote by p_3 the greatest integer such that $p_3/q_3 \le x$ for all $x \in P$. And so on. Now, two alternatives exist: either, for some natural number m, p_m/q_m is the greatest lower bound of P, or else we get an infinite sequence none of whose elements is the greatest lower bound of P. It suffices to consider the second case. Then $\{p_n/q_n\}$ is defined by induction so that $q_{n+1} = 2q_n$ and p_{n+1} is the greatest integer such that $p_{n+1}/q_{n+1} \le x$ for all $x \in P$.

Evidently the sequence $\{p_n/q_n\}$ is bounded and nondecreasing. By 2.12.5 it thus belongs to an equivalence class and represents a real number $a = [\{p_n/q_n\}]$. To prove that a is the greatest lower bound of P suppose that b is a real number such that $a < b \le x$ for all $x \in P$. By 2.12.4, there is a rational number $\varepsilon > 0$ such that $\varepsilon < b - a$. Hence $a + \varepsilon < b$. Let n_0 be the least natural number such that $1/\varepsilon < q_{n_0}$. By 2.12.5, we have $p_{n_0}/q_{n_0} \le a$ and hence

$$\frac{p_{n_0}+1}{q_{n_0}} = \frac{p_{n_0}}{q_{n_0}} + \frac{1}{q_{n_0}} \le \frac{p_{n_0}}{q_{n_0}} + \varepsilon \le a + \varepsilon < b.$$

Therefore $(p_{n_0}+1)/q_{n_0} \le x$ for all $x \in P$, contrary to the definition of p_{n_0}. This proves that a is the greatest number satisfying $a \le x$ for all $x \in P$.

Exercises 2.12

1. Prove that $\{1/n\} \sim \{1/n\}$.

2. Prove that

$$\left\{\frac{1}{n}\right\} \sim \left\{\frac{n}{n^2+1}\right\}.$$

3. Prove that

$$\left\{\frac{2n+1}{n}\right\} \sim \left\{\frac{4n-3}{2n+1}\right\}.$$

4. Prove that $[\{n\}]$ is empty.

5. Prove that $[\{(-1)^n\}]$ is empty.

6. Prove that addition of real numbers is well defined; that is, $\{a_n\} \sim \{b_n\}$ and $\{c_n\} \sim \{d_n\}$ implies $\{a_n + c_n\} \sim \{b_n + d_n\}$.

7. Prove that multiplication of real numbers is well defined, that is, if $\{a_n\} \sim \{b_n\}$ and $\{c_n\} \sim \{d_n\}$ then $\{a_n c_n\} \sim \{b_n d_n\}$.

8. Prove that the inequality for real numbers if well defined, that is, if $[a_n] \sim [b_n]$ and if there exist a positive rational number ε_a and a natural number K_a such that $a_n > \varepsilon_a$ for all $n > K_a$, then there exist a positive rational number ε_b and a natural number K_b such that $b_n > \varepsilon_b$ for all $n > K_b$.

9. Let a and b be two rational numbers. Let $\alpha = [\{a,a,a,\ldots\}]$, $\beta = [\{b,b,b,\ldots\}]$, and $\gamma = [\{a+b, a+b, a+b, \ldots\}]$. Prove that $\alpha + \beta = \gamma$.

10. Let a and b be two rational numbers. Let $\alpha = [\{a,a,a,\ldots\}]$, $\beta = [\{b,b,b,\ldots\}]$, and $\gamma = [\{ab, ab, ab, \ldots\}]$. Prove that $\alpha\beta = \gamma$.

11. Let a be a rational number and let $\alpha = [\{a, a, a, \ldots\}]$. Prove that $a > 0$ if and only if $\alpha > 0$.

12. Explain the importance of the properties described in Exercises 9, 10, and 11.

13. Prove that if $a_1, a_2, \ldots$ and b are rational numbers such that

$$a_n \ge a_{n+1} > b \quad \text{for all } n \in \mathbb{N},$$

then the sequence $\{a_n\}$ represents a real number and $a_m \ge [\{a_n\}]$ for each $m \in \mathbb{N}$.

14. Prove that $\{a_n\} \sim \{(-1)^n a_n\}$ if and only if $[\{a_n\}] = 0$.

15. True or false? If $a_n > 0$ for all $n \in \mathbb{N}$, then $[\{a_n\}] > 0$.

16. Define

$$a_1 = 1; \quad \text{and} \quad a_{n+1} = \frac{3a_n + 4}{2a_n + 3} \quad \text{for } n = 1, 2, 3, \ldots.$$

Prove that $\{a_n\}$ represents a real number.

17. Prove that $[\{a_n\}] > 0$ if and only if there exist a sequence $\{b_n\}$ and a positive rational number ε such that $\{a_n\} \sim \{b_n\}$ and $b_n > \varepsilon$ for all $n \in \mathbb{N}$.

2.13 ON CAUCHY SEQUENCES

A sequence $\{a_n\}$ is called a *Cauchy sequence* if $\{a_n\} \sim \{a_n\}$; that is, for any positive ε there exists a number n_0 such that

$$-\varepsilon < a_m - a_n < \varepsilon \quad \text{for all } m, n \geq n_0.$$

According to 2.12.1 all sequences in equivalence classes are Cauchy sequences.

—Here an explanation is needed. Cauchy sequences are introduced in all books where the method of equivalence classes is used in the construction of real numbers. However in our presentation we did not need the concept of a Cauchy sequence at all.

—Why?

—Because our definition of equivalence is simpler than that usually given in books. Reflexivity, that is, $\{a_n\} \sim \{a_n\}$, is not postulated.

—Why does it help?

—The traditional method consists of two steps. First one distinguishes some sequences, called Cauchy sequences. Then the equivalence relation is defined only for Cauchy sequences. In our approach, the first step becomes superfluous. All the "bad" sequences like $\{1, 2, 3, \ldots\}$ drop out automatically.

Cauchy sequences are named after the great French mathematician Augustin-Louis Cauchy (1789–1857).

Exercises 2.13

1. Prove that $\{n/(n + 1)\}$ is a Cauchy sequence.

2. Show that $\{(n^2 + 1)/(n + 1)\}$ is not a Cauchy sequence.

3. Prove that every Cauchy sequence is bounded.

4. Let $\{a_n\}$ and $\{b_n\}$ be Cauchy sequences. Show that $\{a_n + b_n\}$ and $\{a_n b_n\}$ are Cauchy sequences.

5. Let $\{a_n\}$ and $\{b_n\}$ be sequences of rational numbers. Prove that the following two conditions are equivalent:

    (a) $\{a_n\} \sim \{b_n\}$;

    (b) $\{a_1, b_1, a_2, b_2, a_3, b_3, \ldots\}$ is a Cauchy sequence.

6.   A sequence $\{a_n\}$ is called a *subsequence* of $\{b_n\}$ if $a_n = b_{p_n}$ for some strictly increasing sequence of natural numbers $p_1, p_2, p_3, \ldots$. Prove that a subsequence of a Cauchy sequence is a Cauchy sequence.

2.14 COUNTABLE SETS

Suppose we have two boxes of matches: one with white matches and the other one with red. Is there a way, for a person who cannot count, to find out which box contains more matches?

Let us take one match from the first box and one from the second and put them aside. From the remaining matches again take one match from each box and put them aside too. Continuing doing so we will either exhaust both boxes simultaneously or one of the boxes will become empty when in the second box some matches still remain. The first case occurs if and only if the original number of matches was the same in both boxes. In the second case the box that becomes empty had fewer matches. Note that we do not have to count the matches.

Using the described idea, we can check whether two sets have the same quantity of elements, without counting them. Two sets are called *equipotent* if to every element from the first set we can assign exactly one element from the second set in such a way that each element of the second set is assigned to exactly one element of the first set. Then we say that there is a *one-to-one correspondence* between those sets.

Obviously, every set is equipotent with itself. Moreover, if a set A is equipotent with a set B, then B is equipotent with A. Equipotency is also transitive; that is, if a set A is equipotent with a set B and B is equipotent with a set C, then A is equipotent with C. Thus equipotency is an equivalence relation.

In order to establish a correspondence between elements of equipotent sets we do not necessarily need to join elements in pairs; it suffices to give a law which says how the elements should be joined. Here, however, facts may come to light which seem paradoxical at first glance. For instance, the set of all natural numbers is equipotent with the set of all even numbers. To establish a one-to-one correspondence we associate with every natural number the number that is twice as large (and thus even). This is an example of a set that is equipotent with one of its proper subsets. This property can be used as a definition of an infinite set: a set is called *infinite* if it is equipotent with a proper subset of itself. On the other hand, a nonempty set is called *finite* if it is not equipotent with any proper subset of itself. Finite sets can be also defined as follows: a set A is finite if there exists a natural number n such that A is equipotent with $\{1, 2, \ldots, n\}$.

A set that is equipotent with the set of all natural numbers is called *countable*. Elements of such a set can be numbered so that to each element of the set there corresponds a natural number and each natural number corresponds to one and only one element of the set. All elements of a countable set can thus be arranged into an infinite sequence. Conversely, if all elements of a set can be arranged into an infinite sequence such that every element appears in the sequence exactly once, then the set is countable. Hence the union of a countable set and a finite set or another countable set is countable. It is also clear that a subset of a countable set is either countable or finite.

2.14.1 Theorem *The union of a countable family of countable sets is countable.*

Proof. Let $Z_1, Z_2, Z_3, \ldots$ be a sequence of countable sets. The elements of each set Z_n can be arranged into an infinite sequence

$$a_{n,1}, \ a_{n,2}, \ a_{n,3}, \ldots.$$

Consequently the elements of the union can be arranged into an infinite matrix

$$
\begin{array}{llll}
a_{1,1} & a_{1,2} & a_{1,3} \cdots \\
a_{2,1} & a_{2,2} & a_{2,3} \cdots \\
a_{3,1} & a_{3,2} & a_{3,3} \cdots \\
\cdots \\
a_{n,1} & a_{n,2} & a_{n,3} \cdots \\
\cdots
\end{array}
$$

Now let us split the entries of the matrix into finite groups such that the nth group consists of all elements $a_{i,j}$ such that $i + j = n + 1$. Thus in the nth group we have

$$a_{n,1}, \ a_{n-1,2}, \ a_{n-2,3}, \ldots, a_{2,n-1}, a_{1,n}.$$

They form the nth diagonal of the matrix.

By writing down the first diagonal, then the second diagonal, and so on, we arrange all entries of the matrix into a sequence:

$$a_{1,1}, \ a_{2,1}, \ a_{1,2}, \ a_{3,1}, \ a_{2,2}, \ a_{1,3}, \ldots.$$

In this sequence some elements of the union appear several times if some of the sets Z_n have elements in common. In such a case we have to remove successively each element that is equal to one of the preceding elements. In this way we obtain an arrangement of the union in a sequence, proving the theorem.

2.14.2 Theorem *The set of all rational numbers is countable.*

Proof. Let Z_n denote the set of all quotients of natural numbers whose denominator is n. Thus Z_n consists of the numbers $1/n, 2/n, 3/n, \ldots$. The union of all the sets Z_n is countable, according to the preceding theorem. Thus the set of all positive rational numbers is countable. Similarly the set of all negative rational numbers is countable. If we add to them 0, we obtain the set of all rational numbers.

By an *uncountable* set we mean a nonempty set that is neither finite nor countable. An uncountable set cannot be arranged into a sequence.

2.14.3 Theorem *The set of all real numbers is uncountable.*

We shall show in Section 4.9 that every real number between 0 and 1 can be uniquely represented in the form of an infinite sequence of digits

(1) $$0.d_1 d_2 d_3 \ldots$$

such that there are arbitrarily large indices n for which $d_n \neq 9$. (This is called an expansion into a *decimal fraction*.) On the other hand, any sequence of that form represents a real number between 0 and 1. These facts will be used in the following proof.

Proof of 2.14.3. Suppose that the set of all real numbers x with $0 < x < 1$ can be arranged into a sequence $a_1, a_2, a_3, \ldots$, where

$$a_n = 0.d_{n,1} d_{n,2} d_{n,3} \ldots$$

is the expansion of a_n into a decimal fraction. We construct a new decimal fraction

(2) $$0.e_1 e_2 e_3 \ldots$$

such that $e_n \neq d_{n,n}$ and $e_n \neq 9$. The sequence (2) represents a number between 0 and 1, though it is not among $a_1, a_2, a_3, \ldots$. Thus the assumption that all real numbers between 0 and 1 can be arranged into a sequence leads to a contradiction. This proves that the set of all real numbers between 0 and 1 is uncountable and all the more the set of all real numbers is uncountable.

Real numbers that are not rational numbers are called *irrational numbers*. Note that Theorems 2.14.2 and 2.14.3 imply that there are more irrational numbers than rational numbers.

Exercises 2.14

1. Is the set of all finite sequences of rational numbers countable?
2. Is the set of all infinite sequences of rational numbers countable?
3. Is the set of all infinite sequences of natural numbers countable?
4. Is the set of all infinite binary sequences countable?
5. A number $\alpha \in \mathbb{R}$ is called *algebraic* if it is a root of a polynomial with integer coefficients; that is, there are integers $\lambda_0, \lambda_1, \ldots, \lambda_n$, not all zero, such that

   $$\lambda_0 \alpha^n + \lambda_1 \alpha^{n-1} + \cdots + \lambda_{n-1} \alpha + \lambda_n = 0.$$

   Is the set of all algebraic numbers countable?
6. Prove that every nonempty interval contains infinitely many rational numbers.
7. Prove that every nonempty interval contains infinitely many irrational numbers.

CHAPTER 3

FUNCTIONS

3.1 THE CONCEPT OF A FUNCTION

The function is a very important tool in describing phenomena of the surrounding world, in which there is dependence between entities. For example, how does distance traveled depend on elapsed time? How does the pressure of a gas depend on its temperature? Such questions suggest that the concept of a function stems naturally from that of variable entities, called *variables* for short. However, the concept of a variable is difficult to define accurately and all attempts to do so seem to obscure the concept of a function rather than clarify it. In other books a function is defined as a correspondence or a rule. But in fact such definitions only replace one word by another. The reader is assumed not to know what a function is but is supposed to know what a correspondence is. It seems that it would be better to say that everybody knows what a function is and therefore we do not have to define it. On the other hand we do not want to base a mathematical theory on such an assumption, because it would sooner or later cause some problems. There are important examples in the history of mathematics of such unfortunate cases. So what do we do if we want to have a precise definition of a function? The modern concept of a function is based on the concept of relation or, equivalently, on the idea of a set of ordered pairs.

A *function* f from a set X to a set Y is a collection of ordered pairs (x, y) such that

(a) $x \in X$ and $y \in Y$;

(b) For every $x \in X$ there exists $y \in Y$ such that $(x, y) \in f$;

(c) If $(x, y_1) \in f$ and $(x, y_2) \in f$, then $y_1 = y_2$.

—Maybe the above formal definition of a function is a very rigorous one but it is rather artificial. We often say that $y = x^2$ or $f(x) = x^2$ is a function. According to the above definition it is not, because $f(x) = x^2$ is not a collection of ordered pairs.

—Strictly speaking this is true. On the other hand it is easy to see that the formula $f(x) = x^2$ describes a collection of pairs: a pair $(x, y) \in f$ if and only if $y = x^2$. Using symbols, we can write $f = \{(x, x^2) : x \in \mathbb{R}\}$.

—I like the usual method better. The formula $f(x) = x^2$ tells me all I need to know and it is simpler. One has to agree that, for example, "$\sin \pi = 0$" looks much better than $(\pi, 0) \in \sin$. What about the ugly appearance of "$(2, 4) \in {}^2$"?

—We are not going to change the way we describe functions. It is important however to know that the concept of a function can be rigorously defined in terms of primitive notions; in this case it is actually the concept of a set because an ordered pair can be defined as $(a, b) = \{\{a\}, \{a, b\}\}$.

On the other hand, if you think of the general notion of a function, the formal definition is in some sense natural and necessary. It is very convenient if a function can be given by a short formula like $f(x) = x^2$ or $f(x) = \sin x$, but what if there is no such formula? If you conduct an experiment in which you want to find a relation between two quantities, then the result of the experiment is a set of pairs. For instance, you drop a small ball from a height h and measure the time t it takes to hit the ground. The data from your experiment are going to constitute a table; for example:

h feet	1	1.5	2	2.5	3	3.5	4	4.5	5
t seconds	0.25	0.31	0.35	0.40	0.43	0.47	0.50	0.53	0.56

This is actually a set of pairs, and thus it represents a function. Of course, at this point we could try to find a formula which would describe the correspondence. Such a formula would amount to an extension of the correspondence and would represent a larger collection of ordered pairs. A table representation of a function is often used. Its main disadvantage is that it can contain only a finite number of entries.

In the formal definition of a function f there are three elements: a set X, a set Y, and a collection of pairs. The set X is called the *domain* of f. The set of all elements of Y that are values of f is called the *range* of f. In other words, $y \in Y$ is an element of the range of f if there is an $x \in X$ such that $f(x) = y$. In order to completely define a function it is not enough to give a formula that describes the correspondence. We need to specify its domain. The range is usually determined by the domain. For instance, if $f(x) = x^2$ and the domain is the set of all real numbers, then the range is the set of all non-negative numbers. If $f(x) = x^2$ and the domain is the set of all numbers that are less than -2, then the range is the set of all numbers that are greater than 4. Instead of saying "f is a function with domain X" we can simply say "f is a function on X." The notation "$f : X \to Y$" is very useful. It means "f is a function with domain X whose range is a subset of Y."

In calculus textbooks we often are asked to find the domain of a function. To be more precise we should ask what is the largest set of real numbers for which the rule for the function is well defined. The problem is more or less an arithmetical one. For example, the function $f(x) = x^2$ is defined for all real numbers; $f(x) = 1/x$ is defined for all numbers except $x = 0$; $f(x) = \sqrt{x}$ is defined for all non-negative numbers. If we are asked to find the range of a function we similarly mean the largest possible set of values.

The most important property of a function is property (c): for every x there exists exactly one y such that $f(x) = y$. For example $f(x) = x^2$ describes a function because every real number has exactly one square. It is possible, however, that the same y is assigned to more than one x: $1^2 = 1$ and $(-1)^2 = 1$.

Instead of saying "the function defined by the formula $f(x) = x^2$" we often say "the function $f(x) = x^2$" or even "the function x^2." The last abbreviation is not quite correct, for x^2 is not a function, but a number associated with x. In cases when doubt could arise, it is better to use another symbol, for example, $\{x^2\}$. Similarly, the function defined by the formula $f(x) = 1/x$ would be denoted by $\{1/x\}$. It may happen that we do not know a general formula for a function or, for some reason, we do not want to use it. The number associated with x can be denoted by $f(x)$ and the function itself by $\{f(x)\}$ or just by f. If we have to use other functions simultaneously we may use other letters like g or h.

A systematic use of braces $\{\}$ to distinguish a function from its values is cumbersome in practice. In the rest of the book, therefore, we will drop the braces if there is no fear of misunderstanding.

Some special functions are traditionally denoted by special symbols. For example, the function that assigns to x the greatest integer less than or equal to x is usually denoted by $f(x) = [x]$. In particular, for every $x \in \mathbb{R}$, we have

$$x - 1 < [x] \le x.$$

In the following chapters we will also discuss functions denoted by ln, sin, cos, tan, cot, arctan, arcsin, and arccos.

Exercises 3.1

1. Can you guess an algebraic formula for the function described approximately in the table given in this section?

2. Find the range of each of the following functions $f : \mathbb{R} \to \mathbb{R}$:
    (a) $f(x) = 1 + x^2$.
    (b) $f(x) = (1 + x)^2$.
    (c) $f(x) = \dfrac{1}{1 + x^2}$.
    (d) $f(x) = 1 + x^3$.
    (e) $f(x) = (1 + x)^3$.
    (f) $f(x) = [x]$.
    (g) $f(x) = [x^2 + 7]$.

3. Which expressions can be used to define a function?
    (a) The bigger of the numbers $x + 7$ and $\frac{1}{2}x$.
    (b) The smaller of the numbers x^2 and $2x$.
    (c) A number whose square equals $3x$.
    (d) A number whose cube equals $x + 1$.
    (e) A number that multiplied by x gives 11.

3.2 GEOMETRIC INTERPRETATION OF NUMBERS AND FUNCTIONS

Formerly mathematical analysis was often based on geometric proofs. In the modern approach geometry is completely eliminated as a proving instrument in analysis. Nevertheless, connecting analysis with geometry is very important, because it provides intuitive interpretation of many ideas and offers precious hints in various investigations.

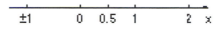

Figure 3.1 A real axis.

In this book geometry will be used only to illustrate the text. It will be treated rather intuitively, without observing strict accuracy.

Numbers can be interpreted geometrically as points on a straight line in such a way that every real number corresponds to one and only one point on the line, and conversely, every point on the line corresponds to exactly one real number. This correspondence can be set up in infinitely many ways. We first choose an arbitrary point on the line and assign to it the number 0. Then we choose another point (traditionally it is a point to the right of the one assigned to 0) and call it 1. The distance from 0 to 1 can be used to assign numbers to the remaining points. If a point is to the right of 0 then we assign to it the positive real number equal to the distance from zero to this point, where the distance from 0 to 1 is the unit. The same rule applies to points to the left of 0 with the only difference being that negative numbers are used. A line for which this is done is called a *real axis* (see Fig. 3.1).

We often use a geometrical way of speaking. For instance, we say "the value of the function f at a point x." This means: "a number assigned by the function f to a number x."

By an *open interval* (a, b) we mean the set of all numbers x such that $a < x < b$. If we include the numbers a and b, then we obtain a closed interval $[a, b]$. Thus the *closed interval* $[a, b]$ is the set of all numbers x such that $a \leq x \leq b$. The numbers a and b are called the *end-points* of the interval; the numbers x satisfying the inequalities $a < x < b$ are called the *interior points* of the interval. Every point in the open interval (a, b) is an interior point.

Open and closed intervals have a simple geometric interpretation. They represent segments between points a and b: an open interval is represented by a segment without ends and a closed interval, with ends (see Fig. 3.2).

We will also use *infinite intervals:* (a, ∞), $(-\infty, b)$, and $(-\infty, \infty)$. The first denotes the set of all real numbers greater than a, the second the set of all numbers less than b, and the third the set of all real numbers. Other intervals often used are the so-called half-open or half-closed intervals. Here are all possible intervals.

$$x \in (a, b) \text{ means } a < x < b;$$
$$x \in (a, b] \text{ means } a < x \leq b;$$
$$x \in [a, b) \text{ means } a \leq x < b;$$
$$x \in [a, b] \text{ means } a \leq x \leq b;$$
$$x \in (a, \infty) \text{ means } x > a;$$
$$x \in [a, \infty) \text{ means } x \geq a;$$

Figure 3.2 Open and closed intervals.

$$x \in (-\infty, b) \quad \text{means } x < b;$$
$$x \in (-\infty, b] \quad \text{means } x \le b;$$
$$(-\infty, \infty) = R.$$

When expressing an interval with left end-point a and right end-point b, we always assume that $a < b$.

Given two nonparallel real axes, we may associate with each point in the plane its projections on the axes. In this way, to each point of the plane there correspond two numbers and, conversely, to each pair of numbers there corresponds a point of the plane. We usually assume that the axes are perpendicular and that the intersection point is 0 on both lines. The intersection is called the *origin*. A plane with two perpendicular axes is called a *coordinate plane* (see Fig. 3.3).

Because a function is a set of pairs of numbers, it can be represented by a set of points in the plane. This set of points is called the *graph* of that function. A graph can be sketched by first constructing a table of the function and then plotting the points on the plane. If the points are dense enough, then we complete the sketch by guessing its shape. We cannot expect great accuracy but it usually gives an idea what the graph looks like. Technicians often use a French curve to complete the sketch. It does not necessarily improve the accuracy, it only gives a better appearance. Nowadays computers can generate graphs of functions with practically arbitrary accuracy.

Figure 3.4 contains a few examples of graphs; as shown, we need to indicate which axis corresponds to which member of the pair (x, y).

Exercise 3.2

1. Sketch graphs of the following functions in the given intervals:

(a) $a(x) = [x^2]$ for x in $[0, 3]$.

(b) $b(x) =$ the distance from x to 3 for x in $[-5, 5]$.

(c) $c(x) =$ the average distance from x to -1 and 1 for x in $[-5, 5]$.

(d) $d(x) = \left[\dfrac{[x] + 1}{3} \right]$ for x in $[0, 10]$.

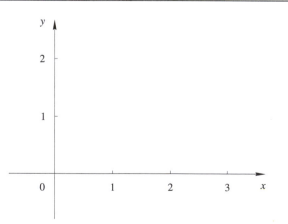

Figure 3.3 A coordinate plane.

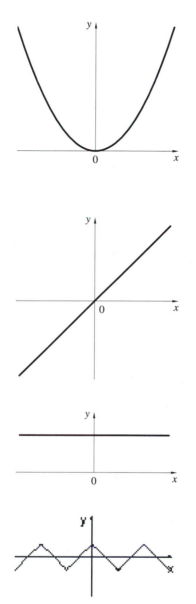

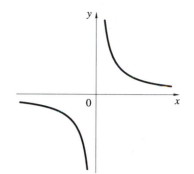

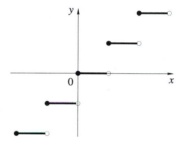

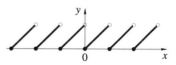

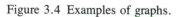

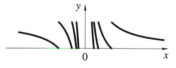

Figure 3.4 Examples of graphs.

3.3 ABSOLUTE VALUE

The *absolute value* of a non-negative number is the number itself, and the absolute value of a negative number is the negative of the number:

$$
(1) \qquad |a| = \begin{cases} a & \text{if } a \geq 0, \\ -a & \text{if } a < 0. \end{cases}
$$

Notice that the absolute value is always a non-negative number. The absolute value of a number different from zero is a positive number. For instance $|5| = 5$, $|0| = 0$, and $|-7| = 7$.

3.3.1 $|a| = |-a|$. *(The sign may be changed under the absolute value.)*

Proof. The assertion is obvious if $a = 0$. If $a > 0$, then $|a| = a$ and $|-a| = -(-a) = a$. If $a < 0$, then $|a| = -a$ and $|-a| = -a$, because $-a > 0$.

3.3.2 $-|a| \leq a \leq |a|$.

Proof. If $a \geq 0$, then $-a \leq a$ and, by (1), we have $-|a| \leq a = |a|$. If $a < 0$, then $a \leq -a$ and, by (1), $-|a| = a \leq |a|$.

3.3.3 *From the inequality*

$$
(2) \qquad -c \leq a \leq c
$$

it follows that

$$
(3) \qquad |a| \leq c,
$$

and, conversely, (3) implies (2).

Proof. Assume first that (2) holds. If $0 \leq a$, then the second inequality in (2) gives $|a| \leq c$; if $a < 0$, then the first inequality in (2) gives $-c \leq -|a|$, thus again $|a| \leq c$.

Assume now that (3) holds. Then we have $-c \leq -|a|$, and hence, by 3.3.2 and (3), we obtain

$$
-c \leq -|a| \leq a \leq |a| \leq c.
$$

3.3.4

$$
|a| - |b| \leq |a + b| \leq |a| + |b|;
$$
$$
|a| - |b| \leq |a - b| \leq |a| + |b|.
$$

Proof. Adding inequalities $-|a| \leq a \leq |a|$ and $-|b| \leq b \leq |b|$ we get

$$
-(|a| + |b|) \leq a + b \leq |a| + |b|.
$$

Hence

$$
(4) \qquad |a + b| \leq |a| + |b|,
$$

by 3.3.3. Replacing b by $-b$ in (4) we get

(5) $|a - b| \le |a| + |b|,$

which implies $|(a + b) - b| \le |a + b| + |b|$, and therefore

(6) $|a| - |b| \le |a + b|.$

Finally, replacing b by $-b$ in (6) we obtain

(7) $|a| - |b| \le |a - b|.$

The inequalities in 3.3.4 are abbreviated forms of (4), (5), (6), and (7).

3.3.5 $|ab| = |a||b|$ and $\left|\dfrac{a}{b}\right| = \dfrac{|a|}{|b|}$ $(b \ne 0).$

(The absolute value of a product equals the product of the absolute values, and the absolute value of a quotient equals the quotient of the absolute values.)

Proof. The equalities are obvious if $a \ge 0$ and $b \ge 0$. In the remaining cases we have

$$|a(-b)| = |a|| - b| \text{ for } a \ge 0,   b < 0;$$
$$|(-a)b| = | - a||b| \text{ for } a < 0,   b \ge 0;$$
$$|(-a)(-b)| = | - a|| - b| \text{ for } a < 0,   b < 0;$$

because all the numbers within the absolute value signs are non-negative. This proves the first equality, by 3.3.1. If $b \ne 0$, then from what we just proved we have $\left|b\dfrac{a}{b}\right| = |b|\left|\dfrac{a}{b}\right|$; that is, $|a| = |b|\left|\dfrac{a}{b}\right|$. This gives the desired equality.

Exercises 3.3

1. Graph the following functions:
    (a) $a(x) = |x|$ on $[-5, 5]$.
    (b) $b(x) = ||x| - 1|$ on $[-5, 5]$.
    (c) $c(x) = |||x| - 1| - 1|$ in $[-5, 5]$.
    (d) $d(x) = \dfrac{|x + 1|}{|x| + 1}$ on $[-5, 5]$.
2. Prove that if $b - 1 < a < b + 1$ then $|a| < |b| + 1$.
3. Graph on the real axis the sets of numbers satisfying the following inequalities:
    (a) $|x - 3| < 2$.
    (b) $|x + 2| \le 3$.
    (c) $0 < |x - \frac{1}{2}| \le 2$.
    (d) $|x - a| < 1$.
    (e) $|x + b| \ge 4$.
    (f) $|x^2 - 4| < \frac{1}{3}$.

4.   Prove that

$$\max\{x, y\} = \frac{1}{2}(x + y + |x - y|) \quad \text{and} \quad \min\{x, y\} = \frac{1}{2}(x + y - |x - y|)$$

for any $x, y \in \mathbb{R}$.

3.4 LIMIT OF A FUNCTION

The function $f(x) = (x - 2)^2$ (see Fig. 3.5) assumes the value 0 at $x = 2$. If x is only a little greater than 2, then the value of the function differs only a little from 0. Let us now consider the function defined by

$$f(x) = \frac{-(x - 2)^2}{|x - 2| + (x - 2)}$$

(see Fig. 3.6). If $x \leq 2$, then $|x - 2| = -(x - 2)$ and the denominator equals 0. Thus the function is not defined at $x = 2$. It is defined only for $x > 2$ because then $|x - 2| = x - 2$ and the value of the function is $\frac{-1}{2}(x - 2)$. Therefore, if x is only a little greater than 2, then the value of the function differs only a little from 0. This shows that the discussed property is not connected with the value of the function at the point 2, but rather with points to the right of that point. If x is close to 2, then the value of the function at x is close to 0. One can also say that values of the function tend to zero when x approaches 2. The following definition expresses this situation in the precise language of analysis.

3.4.1 Definition   We say that the *right-hand limit* of a function f at a point a is k, if for every positive ε there exists a positive number δ such that for every $x \in (a, a + \delta)$ we have $|f(x) - k| < \varepsilon$. We then write

$$\lim_{x \to a+} f(x) = k \quad \text{or} \quad f(a+) = k \,.$$

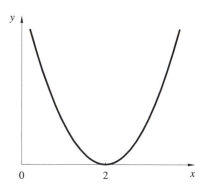

Figure 3.5 Graph of $y = (x - 2)^2$

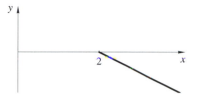

Figure 3.6 Graph of $y = \frac{-(x-2)^2}{|x-2|+x-2}$

—I don't like this definition. It seems to be unnecessarily complicated. I understand the examples before the definition, but it is not easy to see the connection.

—Only for beginners. Probably the difficulty arises from the use of quantifiers.

—What are quantifiers?

—The expressions "for every" and "there exists" are usually called *quantifiers*. In our definition we have three quantifiers: "for every," then "there exists," and finally again "for every." The order of these quantifiers is essential. For ε we can take any positive number, whereas the choice of δ depends on ε. Quantifiers play a very important role in the language of mathematics. Other synonyms are also used. For instance, instead of "for every" one can say "for each," "for all," or "for any." Similarly, "there exists" can be replaced by "there is," "for some," or "for certain." The quantifier "for every" is often omitted. For example, if we say "for $a < x < a + \delta$ we have $|f(x) - k| < \varepsilon$," we actually mean "for every x such that $a < x < a + \delta$ we have $|f(x) - k| < \varepsilon$."

A function tending to 0 may behave in a more complicated manner than in the examples given above. For instance, the right-hand limit at 0 of the function

$$f(x) = x \left[\frac{2}{x} \right] + \frac{x}{2} - 2$$

is 0, and it assumes both positive and negative values in any neighborhood to the right of zero (see Fig. 3.7).

The left-hand limit is defined similarly to the right-hand limit.

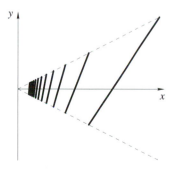

Figure 3.7 Graph of
$y = x \left[\frac{2}{x} \right] + \frac{x}{2} - 2.$

3.4.2 Definition   We say that the *left-hand limit* of a function f at a point a is k, if for every positive ε there exists a positive number δ such that for every $x \in (a - \delta, a)$ we have $|f(x) - k| < \varepsilon$. We write

$$\lim_{x \to a-} f(x) = k \quad \text{or} \quad f(a-) = k.$$

If both the left-hand limit and right-hand limit exist and are equal to k, then we say that the function has the limit k at a. In other words:

3.4.3 Definition   We say that the *limit* of a function f at a point a is k, if for every positive ε there exists a positive number δ such that for every x such that $0 < |x - a| < \delta$ we have $|f(x) - k| < \varepsilon$. We write

$$\lim_{x \to a} f(x) = k.$$

This is also referred to as a *two-sided* limit.

In particular, $\lim_{x \to a} x = a$. If $f(x) = c$ for $a < x < b$, then $\lim_{x \to a+} f(x) = c$, $\lim_{x \to b-} f(x) = c$, and $\lim_{x \to d} f(x) = c$ for $a < d < b$. In other words, a constant function $f(x) = c$ has the limit c at every point.

It is very important to remember that a function need not be defined at a point to have a limit at that point. In the two examples considered above that is exactly the case:

$$f(x) = \frac{(x - 2)^2}{|x - 2| + (x - 2)}$$

is not defined at $x = 2$, but

$$\lim_{x \to 2+} \frac{(x - 2)^2}{|x - 2| + (x - 2)} = 0.$$

Similarly, $f(x) = x \left[\dfrac{2}{x} \right] + \dfrac{x}{2} - 2$ is not defined at $x = 0$, but $\lim_{x \to 0+} \left(x \left[\dfrac{2}{x} \right] + \dfrac{x}{2} - 2 \right) = 0$.

For a function f to have a right-hand limit at a it must be defined on an interval of the form $(a, a + \epsilon)$, where ϵ is a positive number. In other words, an interval of the form $(a, a + \epsilon)$ has to be contained in the domain of f. Similarly, to have a left-hand limit at a, f has to be defined on an interval of the form $(a - \epsilon, a)$, $\epsilon > 0$. Finally, to have a two-sided limit at a, f has to be defined on both intervals $(a - \epsilon, a)$ and $(a, a + \epsilon)$ for some positive ϵ. Of course, being defined on those intervals is not sufficient for existence of a limit. Consider, for example, $f(x) = \dfrac{1}{x} - \left[\dfrac{1}{x} \right]$ at $x = 0$ (see Fig. 3.4). This function is defined for all $x \neq 0$, but it does not have a limit at 0, not even a one-sided limit.

The intervals $(a, a + \epsilon)$ and $(a - \epsilon, a)$ are called a *right neighborhood of a* and a *left neighborhood of a*, respectively. The set $(a, a + \epsilon) \cup (a - \epsilon, a)$ is often called a *punctured neighborhood of a*.

To find the limit of a function at a point we often use the fact that if two functions are identical in a neighborhood of a point a and one of them has the limit k at a, then the other one has the limit k at a. More precisely, if the function f is defined

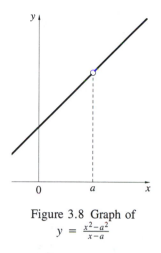

Figure 3.8 Graph of
$$y = \frac{x^2 - a^2}{x - a}$$

on $(a - \delta, a + \delta)$ and has the limit k at a, and h is a function defined on $(a - \delta, a)$ and $(a, a + \delta)$ and equal to f on these intervals, then h has the limit k at a. For instance, the functions

$$f(x) = \frac{x^2 - a^2}{x - a} \quad \text{and} \quad g(x) = x + a$$

(see Fig. 3.8 and Fig. 3.9), are identical everywhere except at $x = a$. Thus, instead of calculating the limit of the first function, we can find the limit of the second one. It is easily seen that

$$\lim_{x \to a} (x + a) = 2a.$$

Hence also

$$\lim_{x \to a} \frac{x^2 - a^2}{x - a} = 2a.$$

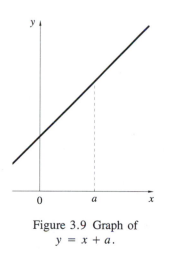

Figure 3.9 Graph of
$$y = x + a.$$

3.4.4 Example   We are going to use Definition 3.4.3 to prove that $\lim_{x \to 1} \frac{1}{x^2 + 1} = \frac{1}{2}$. Let ε be an arbitrary positive number. (We cannot specify ε. We have to present an argument that works for any positive ε.) We need to find a positive number δ such that

$$\left| \frac{1}{x^2 + 1} - \frac{1}{2} \right| < \varepsilon \quad \text{whenever } 0 < |x - 1| < \delta.$$

(Finding δ actually means expressing it in terms of ε. Because it is difficult to guess what δ is going to have that property, we will try to rearrange

$$\left| \frac{1}{x^2 + 1} - \frac{1}{2} \right|$$

to see whether that will give us any hints.) We have

(1) $$\left| \frac{1}{x^2 + 1} - \frac{1}{2} \right| = \left| \frac{1 - x^2}{2(x^2 + 1)} \right| = \left| \frac{(1 - x)(1 + x)}{2(x^2 + 1)} \right| = \left| \frac{1 + x}{2(x^2 + 1)} \right| |x - 1|$$

Now let us find a number α such that

$$\left| \frac{1 + x}{2(x^2 + 1)} \right| \le \alpha.$$

In order to be able to do that we assume that $\delta \le 1$. Consequently, if $0 < |x - 1| < \delta$, then $0 < x < 2$ and hence

(2) $$\left| \frac{1 + x}{2(x^2 + 1)} \right| < \frac{3}{2}.$$

Combining (1) and (2) we obtain

$$\left| \frac{1}{x^2 + 1} - \frac{1}{2} \right| < \frac{3}{2} |x - 1|.$$

Thus, if we define δ to be the minimum of $\frac{2}{3}\varepsilon$ and 1, then

$$\left| \frac{1}{x^2 + 1} - \frac{1}{2} \right| < \frac{3}{2} |x - 1| < \frac{3}{2} \frac{2\varepsilon}{3} = \varepsilon \quad \text{whenever } 0 < |x - 1| < \delta.$$

Now the proof is complete.   ■

Exercises 3.4

1. Use Definitions 3.4.1, 3.4.2, and 3.4.3 to prove the following:

(a) $\lim_{x \to a} x = a$, a is an arbitrary number.

(b) $\lim_{x \to 0} |x| = 0$.

(c) $\lim_{x \to 0+} (x^2 + x + 1) = 1$.

(d) $\lim_{x \to 1-} \frac{1}{x + 1} = \frac{1}{2}$.

(e) $\displaystyle\lim_{x \to -2} \frac{1}{x^2 - 1} = \frac{1}{3}.$

(f) $\displaystyle\lim_{x \to 0+} \frac{x}{|x|} = 1$ and $\displaystyle\lim_{x \to 0-} \frac{x}{|x|} = -1.$

(g) $\displaystyle\lim_{x \to n-} [x] = n - 1$ and $\displaystyle\lim_{x \to n+} [x] = n,$ for any integer n.

(h) $\displaystyle\lim_{x \to 0} |x| \neq 1.$

(i) $\displaystyle\lim_{x \to 0-} (x^2 + x + 1) \neq 2.$

(j) $\displaystyle\lim_{x \to 2+} \frac{\sqrt{x^2 - x + 2}}{\sqrt{x - 2} + x - 2} \neq 0.$

2.   Show that the following limits do not exist:

   (a) $\displaystyle\lim_{x \to 0} \frac{|x|}{x};$

   (b) $\displaystyle\lim_{x \to 1+} \frac{1}{x - 1};$

   (c) $\displaystyle\lim_{x \to -2} [x].$

3.   Let

$$f(x) = \begin{cases} 1 & \text{if } x \text{ is a rational number,} \\ 0 & \text{if } x \text{ is an irrational number.} \end{cases}$$

   Show that $\lim_{x \to a} f(x)$ does not exist at any $a \in \mathbb{R}$.

4.   Let

$$f(x) = \begin{cases} x & \text{if } x \text{ is a rational number,} \\ 0 & \text{if } x \text{ is an irrational number.} \end{cases}$$

   Find all $a \in \mathbb{R}$ for which the limit $\lim_{x \to a} f(x)$ exists.

5.   Prove that if $\lim_{x \to a} f(x) = L$ and $\lim_{x \to a} f(x) = M$ then $L = M$. Explain the meaning of this property. Is the same true for one-sided limits?

6.   Prove that $\lim_{x \to a} f(x) = L$ implies that $\lim_{x \to a} |f(x)| = |L|$. Is the converse true?

7.   Prove that $\lim_{x \to 0} f(x) = L$ if and only if $\lim_{x \to 0} f(-x) = L$. Is the same true for one-sided limits?

3.5 ALGEBRAIC OPERATIONS ON LIMITS

In the following theorem the sign "lim" means any one of the three symbols $\lim_{x \to a+}$, $\lim_{x \to a-}$, or $\lim_{x \to a}$.

3.5.1 Theorem   *If* $\lim f(x) = \alpha$ *and* $\lim g(x) = \beta$, *then*

(a) $\lim(f(x) + g(x)) = \alpha + \beta;$

(b) $\lim(f(x) - g(x)) = \alpha - \beta;$

(c) $\lim f(x)g(x) = \alpha\beta$.

Moreover, if $\beta \neq 0$, then

(d) $\lim \dfrac{f(x)}{g(x)} = \dfrac{\alpha}{\beta}$.

In order to prove the theorem we shall first establish the following.

3.5.2 Lemma   *A function f has the right-hand limit α at a point a if and only if there exists a number $M > 0$ such that for every $\varepsilon > 0$ there exists a positive number δ such that for every $x \in (a, a + \delta)$ we have $|f(x) - \alpha| < M\varepsilon$.*

Before approaching the proof notice that the only difference between the last condition and Definition 3.4.1 is the introduction of the constant M. Obviously, the lemma remains true if the words "right-hand limit" are replaced by "left-hand limit" and the interval $(a, a + \delta)$ is replaced by $(a - \delta, a)$.

Proof.   If M is such that for every $\varepsilon > 0$ there is $\delta > 0$ such that for every $x \in (a, a + \delta)$ we have $|f(x) - \alpha| < M\varepsilon$, then, in particular, it is true for $\dfrac{\varepsilon}{M}$ instead of ε, and hence $|f(x) - \alpha| < M\dfrac{\varepsilon}{M} = \varepsilon$. The converse is even easier, because in order to show that such a number M exists it suffices to take $M = 1$.

Proof of 3.5.1.   We have

(1) $\qquad\qquad |f(x) + g(x) - (\alpha + \beta)| \leq |f(x) - \alpha| + |g(x) - \beta|$.

For every $\varepsilon > 0$ there exist numbers $\delta_1 > 0$ and $\delta_2 > 0$ such that

(2) $|f(x) - \alpha| < \varepsilon$ for $a < x < a + \delta_1, |g(x) - \beta| < \varepsilon$ for $a < x < a + \delta_2$.

Let δ denote the lesser of the numbers δ_1 and δ_2. Then, by (1), we have

$$|f(x) + g(x) - (\alpha + \beta)| < 2\varepsilon \quad \text{for} \quad a < x < a + \delta.$$

Thus the function $f + g$ has the right-hand limit at a equal to $\alpha + \beta$, according to 3.5.2 (with $M = 2$).

Similarly

$$|f(x) - g(x) - (\alpha - \beta)| \leq |f(x) - \alpha| + |g(x) - \beta|$$

and the rest of the proof follows as before.

The proofs of (c) and (d) are slightly more difficult. First notice that

$$f(x)g(x) - \alpha\beta = (f(x) - \alpha)g(x) + \alpha(g(x) - \beta).$$

There is a number $\delta_0 > 0$ such that $|g(x) - \beta| < 1$ for $a < x < a + \delta_0$, and thus

$$|g(x)| < 1 + |\beta|.$$

Hence

$$|f(x)g(x) - \alpha\beta| \leq \varepsilon(1 + |\beta|) + |\alpha|\varepsilon = (1 + |\beta| + |\alpha|)\varepsilon$$

for $a < x < a + \delta$, where δ is the least of the numbers δ_0, δ_1, and δ_2 (where δ_1 and δ_2 are such that (2) holds).

To prove (d) we write

$$\frac{f(x)}{g(x)} - \frac{\alpha}{\beta} = \frac{\beta(f(x) - \alpha) - \alpha(g(x) - \beta)}{\beta g(x)}.$$

There is a number δ_0 such that $|g(x) - \beta| < |\beta|/2$, and thus $|g(x)| \geq |\beta|/2$ for $a < x < a + \delta_0$. Hence

$$\left| \frac{f(x)}{g(x)} - \frac{\alpha}{\beta} \right| < \frac{|\beta|\varepsilon + |\alpha|\varepsilon}{\frac{1}{2}\beta^2} = \frac{2(|\beta| + |\alpha|)}{\beta^2} \varepsilon$$

for $a < x < a + \delta$, where δ is the least of δ_0, δ_1, and δ_2.

For the left-hand limit, the proof is the same: we only have to replace the inequalities $a < x < a + \delta$ by $a - \delta < x < a$. Because Theorem 3.5.1 is true for left-hand and right-hand limits, it is clearly true for the two-sided limit when both one-sided limits exist and are equal.

Theorem 3.5.1 is not only of theoretical but also of practical importance. It allows us to find limits of more complicated functions.

3.5.3 Example

$$\lim_{x \to 1} \frac{2x}{x^2 + x - 1} = 2.$$

Because $\lim_{x \to 1} x = 1$, we have $\lim_{x \to 1} x^2 = 1$, by (c), and $\lim_{x \to 1} x^2 + x = 2$, by (a). Thus $\lim_{x \to 1} x^2 + x - 1 = 1$, by (b). Since $\lim_{x \to 1} 2 = 2$, we have $\lim_{x \to 1} 2x = 2$, by (c). Finally, by (d), we obtain the desired result. ∎

3.5.4 Example

$$\lim_{h \to 0} \frac{\frac{1}{x + h} - \frac{1}{x}}{h} = -\frac{1}{x^2}.$$

In this example the letter h plays the same role as x in the first example and x is an arbitrary constant number different from 0. For $h \neq 0$, we have

$$\frac{\frac{1}{x + h} - \frac{1}{x}}{h} = \frac{-1}{x(x + h)}.$$

Applying Theorem 3.5.1 we easily find that the function on the right-hand side has the limit $-1/x^2$ at $h = 0$. Thus the same limit is assigned to the function on the left-hand side. ∎

Exercises 3.5

1. Find the following limits:

   (a) $\lim\limits_{x \to 1} \dfrac{x^3 + x}{x^4 + x^2 + 1}$; (b) $\lim\limits_{x \to -1} \dfrac{x^2 + x}{x + 1}$;

   (c) $\lim\limits_{x \to 2} \dfrac{x^2 - 4}{x + 2}$; (d) $\lim\limits_{x \to 0} \dfrac{x^2}{|x|}$.

2. Find the following limits:

   (a) $\lim\limits_{h \to 0} \dfrac{(x + h)^2 - x^2}{h}$; (b) $\lim\limits_{h \to 0} \dfrac{\frac{1}{(x + h)^2} - \frac{1}{x^2}}{h}$;

   (c) $\lim\limits_{h \to 0-} \dfrac{|x + h| - |x|}{h}$; (d) $\lim\limits_{h \to 0+} \dfrac{|x + h| - |x|}{h}$.

3. Use the Induction Principle to prove that if the limits $\lim_{x \to a} f_1(x)$, $\lim_{x \to a} f_2(x)$, $\ldots$, $\lim_{x \to a} f_n(x)$ exist then the limit $\lim_{x \to a}(f_1(x) + f_2(x) + \ldots + f_n(x))$ exists, and that

$$\lim_{x \to a}(f_1(x) + f_2(x) + \ldots + f_n(x)) = \lim_{x \to a} f_1(x) + \lim_{x \to a} f_2(x) + \ldots + \lim_{x \to a} f_n(x).$$

4. Prove that if the limit $\lim_{x \to a} f(x)$ exists then $\lim_{x \to a}(f(x))^3$ exists and is given by the equality

$$\lim_{x \to a}(f(x))^3 = (\lim_{x \to a} f(x))^3.$$

3.6 LIMITS AND INEQUALITIES

In the preceding section we investigated the relationship of the limit to four algebraic operations. Now we shall prove two theorems that link the limit with inequalities. As before, the sign "lim" can be replaced by any of the symbols $\lim_{x \to a+}$, $\lim_{x \to a-}$, or $\lim_{x \to a}$.

3.6.1 Theorem (on preserving inequality) *If $f(x) \le g(x)$, then $\lim f(x) \le \lim g(x)$ whenever the limits exist.*

Proof.   Assume that

$$\lim_{x \to a+} f(x) - \lim_{x \to a+} g(x) > 0.$$

Then by Theorem 3.5.1(b),

$$\lim_{x \to a+} (f(x) - g(x)) = k > 0$$

and there exists a number $\delta > 0$ such that $|f(x) - g(x) - k| < k/2$ for $a < x < a + \delta$. Hence $k/2 < f(x) - g(x) < 3k/2$ for $a < x < a + \delta$. Thus $f(x) - g(x) > 0$, contrary to the hypothesis. The proof for the left-hand limit is obtained from the above by replacing $a < x < a + \delta$ by $a - \delta < x < a$.

It is important to remember that in Theorem 3.6.1 the weak inequality cannot be replaced by the strong inequality. For instance, if $f(x) = -x$ and $g(x) = x$, then $f(x) < g(x)$ for $x > 0$, but in spite of this we have $\lim_{x \to 0+} f(x) = \lim_{x \to 0+} g(x) = 0$.

3.6.2 Theorem (on three functions)   *If* $f(x) \le g(x) \le h(x)$ *and* $\lim f(x) = \lim h(x) = k$, *then also* $\lim g(x) = k$.

Proof.   It may seem that 3.6.2 follows at once from 3.6.1. However, the existence of $\lim g(x)$ must be proved. Since

$$f(x) - k \le g(x) - k \le h(x) - k$$

and for every $\varepsilon > 0$ there exist numbers δ_1 and δ_2 such that

$$f(x) - k > -\varepsilon \quad \text{for} \quad a < x < a + \delta_1,$$
$$h(x) - k < \varepsilon \quad \text{for} \quad a < x < a + \delta_2,$$

we have

$$-\varepsilon < g(x) - k < \varepsilon \quad \text{for} \quad a < x < a + \delta,$$

where δ is the smaller of the numbers δ_1 and δ_2. This proves 3.6.2 for the right-hand limit. We extend the result to other types of limits in the usual way.

Exercises 3.6 ————————————————————————————————

1. Show that

$$\lim_{x \to 0+} \left(1 - x \left[\frac{1}{x} \right] \right) = 0.$$

2. Show that

$$\lim_{x \to 0+} \left(x \left[\frac{2}{x} \right] + \frac{x}{2} - 2 \right) = 0.$$

3. Find $\lim_{x \to 0} f(x)$ if

$$f(x) = \begin{cases} 1 + x^2 & \text{if } x \text{ is a rational number,} \\ 1 - x^3 & \text{if } x \text{ is an irrational number.} \end{cases}$$

4. Prove that if $a \le f(x) \le b$ and $\lim_{x \to x_0} f(x)$ exists, then $a \le \lim_{x \to x_0} f(x) \le b$.

5. Let $f : \mathbb{R} \to \mathbb{R}$ satisfy the *Cauchy functional equation*

$$f(x + y) = f(x) + f(y) \quad \text{for all } x, y \in \mathbb{R}.$$

Prove that if f has a limit at 0, it has a limit at every other point.

6. True or false? If $f(x) \leq g(x)$ and $\lim g(x)$ exists, then $\lim f(x)$ exists and $\lim f(x) \leq \lim g(x)$.

3.7 CONTINUITY OF A FUNCTION

Let $f(x) = x^2 + x - 1$. In Example 3.5.3 we proved, using Theorem 3.5.1, that $\lim_{x \to 1} f(x) = 1$. The same number can be obtained in a much simpler way by evaluating the function at $x = 1$: $f(1) = 1^2 + 1 - 1 = 1$. Is there then any difference between the value of a function at a point x_0 and the limit of that function at x_0? Or are these just an easy way and a hard way of solving the same problem?

The first important difference between the limit at a point and the value at a point is that, as we have seen in Section 3.4, the function need not be defined at a point to have a limit there. Conversely, a function defined at a point does not have to have a limit at that point. Consider, for example, the function defined as $f(x) = 1/x$ for $x \neq 0$, and $f(0) = 0$. The value of a function at a point is a property of the function at that point, while the limit of the function is a property of the function in a neighborhood of that point. The equality of the value and the limit is a desirable property called *continuity*. The following definitions treat this notion in more detail.

3.7.1 Definition A function f is called *right-hand continuous* at a point x_0 if $\lim_{x \to x_0+} f(x) = f(x_0)$. A function f is called *left-hand continuous* at a point x_0 if $\lim_{x \to x_0-} f(x) = f(x_0)$.

The function $f(x) = x$ (see Fig. 3.10) is right-hand and left-hand continuous at each point x_0, because $\lim_{x \to x_0+} x = x_0$ and $\lim_{x \to x_0-} x = x_0$. The function $f(x) = [x]$ (see Fig. 3.11) is right-hand continuous at each point, but it is not left-hand continuous at integral points. The function $f(x) = |x|/x$ (see Fig. 3.12) is neither left-hand continuous nor right-hand continuous at zero, because it is not defined there. If we additionally define $f(0) = 1$ or $f(0) = -1$, then it becomes right-hand or left-hand continuous, respectively.

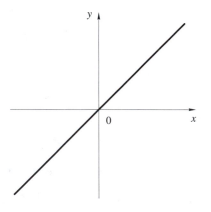

Figure 3.10 Graph of $y = x$

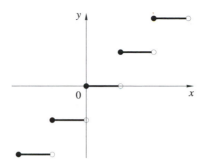

Figure 3.11 Graph of $y = [x]$.

3.7.2 Definition   If a function f is right-hand and left-hand continuous at a point x_0, then we say that f is *continuous* at x_0; that is, f is continuous at x_0 if $\lim_{x \to x_0} f(x) = f(x_0)$. If f is continuous at every point, we simply say that f is continuous.

The functions $f(x) = c$ (a constant) and $f(x) = x$ are continuous at every point. The function $f(x) = (x^2 - 4)/(x - 2)$ is not continuous at $x = 2$ because it is not defined at that point. However, we can make the function continuous by defining $f(2) = 4$. On the other hand the function $f(x) = |x|/x$ (see Fig. 3.12) cannot be made continuous at zero, because the limit at that point does not exist; the left-hand limit is different from the right-hand limit.

3.7.3 Theorem   *If two functions f and g are continuous at x_0, then their sum, difference, and product are continuous at x_0. If, moreover, $g(x_0) \neq 0$, then the quotient f/g is continuous at x_0. The same is true for right-hand and left-hand continuity.*

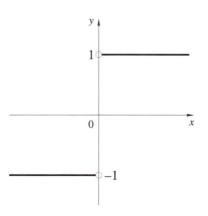

Figure 3.12 Graph of $y = |x|/x$.

Proof.   Assume that f and g are right-hand continuous; that is,

$$\lim_{x \to x_0+} f(x) = f(x_0) \quad \text{and} \quad \lim_{x \to x_0+} g(x) = g(x_0).$$

Then in view of 3.5.1 we have

$$\lim_{x \to x_0+} (f(x) + g(x)) = f(x_0) + g(x_0),$$

$$\lim_{x \to x_0+} (f(x) - g(x)) = f(x_0) - g(x_0),$$

$$\lim_{x \to x_0+} f(x)g(x) = f(x_0)g(x_0),$$

and, if $g(x_0) \neq 0$,

$$\lim_{x \to x_0+} \frac{f(x)}{g(x)} = \frac{f(x_0)}{g(x_0)}.$$

The above proves the theorem for right-hand continuity. To obtain the proof for left-hand continuity it suffices to replace x_0+ by x_0-. Combining both results we obtain the assertion for continuity.

Exercises 3.7

1. Let $f(x) = (x^2 + x - 6)/(x - 2)$. How should f be defined at $x = 2$ so that f will be a continuous function?

2. Prove that the function $f(x) = [x]$ is right-hand continuous at each point, but that it is not left-hand continuous at integral points.

3. Prove that a function f is continuous at x_0 if and only if for every $\varepsilon > 0$ there exists $\delta > 0$ such that $|f(x) - f(x_0)| < \varepsilon$ whenever $|x - x_0| < \delta$.

4. Find an example of a function $f : \mathbb{R} \to \mathbb{R}$ discontinuous at every point such that f^2 is continuous at every point.

5. Show that if f is continuous at a point, then $|f|$ is continuous at that point.

6. Let f and g be continuous at x_0. Show that the functions

$$M(x) = \max\{f(x), g(x)\} \quad \text{and} \quad m(x) = \min\{f(x), g(x)\}$$

are continuous at x_0.

7. Prove that the function

$$f(x) = \begin{cases} 1 & \text{if } x \text{ is a rational number,} \\ 0 & \text{if } x \text{ is an irrational number,} \end{cases}$$

is discontinuous at every point.

8. Prove that the function

$$f(x) = \begin{cases} x & \text{if } x \text{ is a rational number,} \\ 0 & \text{if } x \text{ is an irrational number,} \end{cases}$$

is discontinuous at every point except 0.

9. Prove that *Dirichlet's function,*

$$f(x) = \begin{cases} \frac{1}{q} & \text{if } x = \frac{p}{q} \text{ is a rational number in lowest terms,} \\ 0 & \text{if } x \text{ is an irrational number,} \end{cases}$$

is continuous at every irrational point and discontinuous at every rational point.

10. Find a function continuous at two distinct points a and b and discontinuous everywhere else.

11. Decide whether the following are possible:

    (a) f is continuous at x_0, g is not continuous at x_0, and $f + g$ is continuous at x_0;

    (b) f and g are not continuous at x_0, but $f + g$ is continuous at x_0;

    (c) f is continuous at x_0, g is not continuous at x_0, and fg is continuous at x_0;

    (d) f and g are not continuous at x_0, but fg is continuous at x_0.

12. Let f be a continuous function on $[a, b]$. Prove that if $f(x) = 0$ for every rational x in $[a, b]$ then $f(x) = 0$ for every x in $[a, b]$.

13. Let f and g be two continuous functions on $[a, b]$. Prove that if f and g are equal at rational points of $[a, b]$, then f and g are identical on $[a, b]$.

14. Find all continuous functions $f : \mathbb{R} \to \mathbb{R}$ satisfying the Cauchy functional equation

$$f(x + y) = f(x) + f(y) \quad \text{for all } x, y \in \mathbb{R}.$$

3.8 POWERS

3.8.1 Definition   For natural numbers n the *powers* a^n of a real number a are defined by induction:

$$a^1 = a; \quad a^{n+1} = a \cdot a^n \quad \text{for } n \in \mathbb{N}.$$

Thus we have $a^2 = a \cdot a$, $a^3 = a \cdot a^2 = a \cdot a \cdot a$, $a^4 = a \cdot a^3 = a \cdot a \cdot a \cdot a$, and so on.

3.8.2 Theorem   *The function $f(x) = x^n$ ($n \in \mathbb{N}$) is continuous at every point.*

Proof.   The assertion is true for $n = 1$, because the function $f(x) = x$ is continuous. If the function $f(x) = x^n$ is continuous, then by Theorem 3.7.3 the function $f(x) = x x^n = x^{n+1}$ is also continuous. Thus, Theorem 3.8.2 follows by the induction principle.

The graphs of $f(x) = x^n$ for $n = 1, 2, 3, 4$ are presented in Fig. 3.13.

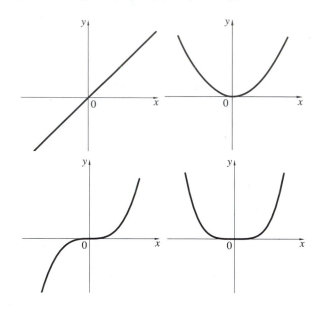

Figure 3.13 Graphs of $y = x$, $y = x^2$, $y = x^3$, and $y = x^4$.

The power x^n can easily be extended to all integers by letting

$$x^0 = 1; \qquad x^{-n} = \frac{1}{x^n} \quad \text{for } n \in \mathbb{N} \quad \text{and } x \neq 0.$$

Exercises 3.8

1. Prove that if $a > 0$, then $a^n > 0$ for any integer n. Is the same true for $a < 0$?
2. Show that all polynomial functions are continuous at every point.
3. A function f is called a *rational function* if it can be written in the form $f(x) = P(x)/Q(x)$, where P and Q are polynomials. Show that every rational function is continuous at every point of its domain.
4. Prove that if f is continuous at a then, for any $n \in \mathbb{N}$, the function $g(x) = (f(x))^n$ is continuous at a.

3.9 FUNCTIONS CONTINUOUS ON AN INTERVAL

We say that a function is continuous on an open interval (a, b) if it is continuous at each point of that interval. We say that a function is continuous on a closed interval $[a, b]$ if it is continuous at every point of the open interval (a, b) and, moreover, it is right-hand continuous at a and left-hand continuous at b.

For example, the function

$$f(x) = \frac{1}{x(3-x)}$$

(see Fig. 3.14) is continuous on $(0, 3)$ but not continuous on $[0, 3]$, because it is not defined for $x = 0$ and $x = 3$. Moreover, it is not possible to define this function at 0 and 3 to make it continuous on $[0, 3]$. The function is continuous on $[1, 2]$, or on every interval $[\varepsilon, 3 - \delta]$ where ε and δ are arbitrary positive numbers such that $\varepsilon < 3 - \delta$.

In general, if f is continuous on an interval, then it is continuous on any subinterval.

3.9.1 Theorem   *If a function f is right-hand continuous at a point a and $f(a) < k$ (or $f(a) > k$), then $f(x) < k$ (or $f(x) > k$) for every x in some interval $[a, a + \delta]$, $\delta > 0$. Similarly, if a function is left-hand continuous at a point a and $f(a) < k$ (or $f(a) > k$), then $f(x) < k$ (or $f(x) > k$) for every x in some interval $[a - \delta, a]$, $\delta > 0$.*

The above properties have a simple geometric interpretation; see Fig. 3.15 and Fig. 3.16.

Proof of 3.9.1.   If f is right-hand continuous at a and $f(a) < k$, then for every $\varepsilon > 0$ there exists a number $\delta > 0$ such that $|f(x) - f(a)| < \varepsilon$ for $a < x < a + \delta$. In particular, if $\varepsilon < k - f(a)$ we have $|f(x) - f(a)| < k - f(a)$, and hence $f(x) - f(a) < k - f(a)$ and $f(x) < k$ for $a < x < a + \delta$. If f is left-hand continuous, the proof is similar, but one has to replace inequalities $a < x < a + \delta$ by $a - \delta < x < a$ and modify properly the remaining part of the proof.

If a function is continuous at a point a, then it is right-hand continuous and left-hand continuous at that point. Thus we have the following corollary:

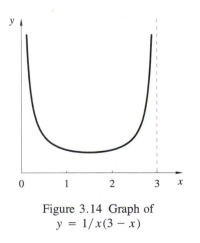

Figure 3.14 Graph of
$y = 1/x(3 - x)$

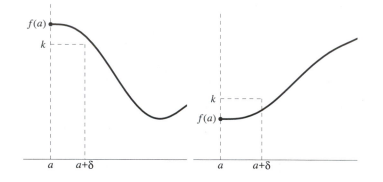

Figure 3.15 Illustration of Theorem 3.9.1 for right-hand continuity.

3.9.2 Corollary   *If a function is continuous at a point a and $f(a) < k$ (or $f(a) > k$), then $f(x) < k$ (or $f(x) > k$) for every x in some interval $[a - \delta, a + \delta]$, $\delta > 0$.*

The graph of a function continuous on an interval can be drawn without lifting the pencil. Such an intuitive understanding of continuity allows us to foresee the following theorem. This is one of the most important properties of continuous functions.

3.9.3 Theorem   *If f is continuous on an interval $[a, b]$ and such that $f(a) < 0$ and $f(b) > 0$, then there exists a point x_0 in (a, b) such that $f(x_0) = 0$. (See Fig. 3.17.)*

Proof.   Let x_0 be the least upper bound of the set Z of all numbers of $x \in [a, b]$ such that $f(x) < 0$. The set is not empty because $a \in Z$. In view of 3.9.1, we have $a < x_0 < b$. We are going to prove that $f(x_0) = 0$. Suppose that $f(x_0) < 0$. Then,

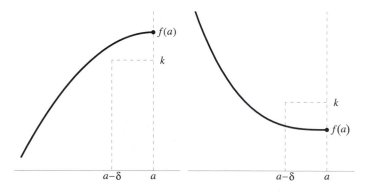

Figure 3.16 Illustration of Theorem 3.9.1. for left-hand continuity.

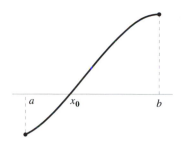

Figure 3.17 Illustration of the
intermediate value property.

by 3.9.2, $f(x) < \frac{1}{2}f(x_0)$ for every x in some interval $[x_0 - \delta, x_0 + \delta]$, and thus x_0 cannot be the least upper bound of Z. Similarly, if $f(x_0) > 0$, then $f(x) > \frac{1}{2}f(x_0)$ for every x in some interval $[x_0 - \delta, x_0 + \delta]$, contradicting the definition of x_0. Therefore $f(x_0) = 0$.

The function $f(x) = x^5 + x - 1$ is continuous on the interval $[0, 1]$ and assumes values 1 and -1 at its ends. Therefore there exists a number $x_0 \in (-1, 1)$ satisfying the equation $x_0^5 + x_0 - 1 = 0$. Theorem 3.9.2 ensures the existence of a solution of some equations, but it does not help to find the solution. Finding a solution might be much more difficult.

An easy corollary of Theorem 3.9.3 is the following:

3.9.4 Theorem (intermediate value theorem)   *A function that is continuous on a closed interval assumes on its interior all values between the initial and the final values. More precisely, if f is continuous on $[a, b]$ and $f(a) \neq f(b)$ then for every number y_0 between $f(a)$ and $f(b)$ there exists a number $x_0 \in (a, b)$ such that $f(x_0) = y_0$.*

Proof.   If $f(a) < y_0 < f(b)$, then the function $g(x) = f(x) - y_0$ is negative at a and positive at b. Thus there exists $x_0 \in (a, b)$ such that $g(x_0) = 0$. Consequently, $f(x_0) = y_0$. If $f(b) < f(a)$, the proof is similar.

Exercises 3.9

1. Prove Theorem 3.9.1 for left-hand continuous functions.

2. Show that the equation $x^5 - 3x^3 - x + 1 = 0$ has at least one solution in the interval $[0, 2]$.

3. Let $f : [a, b] \rightarrow [a, b]$ be a continuous function. Prove that there exists $c \in [a, b]$ such that $f(c) = c$.

4. Let f and g be continuous functions on $[a, b]$. Show that if $f(a) \leq g(a)$ and $f(b) \geq g(b)$ then there exists $c \in [a, b]$ such that $f(c) = g(c)$.

5. Let $f : [a, b] \to \mathbb{Q}$ be a continuous function ($\mathbb{Q}$ denotes the set of rational numbers). Prove that f is a constant function.

6. Show that the range of a continuous function on $[a, b]$ is either a one-element set or an uncountable set.

3.10 BOUNDED FUNCTIONS

We introduce the following definition:

3.10.1 Definition We say that a function f is *bounded* on an interval if there exists a constant M such that $|f(x)| < M$ for all points x of that interval.

The function $f(x) = (2/x) - 1$ (see Fig. 3.18) is bounded on the interval $(1, 3)$, because $|f(x)| < 1$ for all points in $(1, 3)$. However, it is not bounded on $(0, 3)$, because for any number $M > 0$ we have $f(2/(M + 2)) = M + 1$, and $2/(M + 2) \in (0, 3)$ for all $M \in \mathbb{R}$.

3.10.2 Theorem *A function continuous on a closed bounded interval is bounded on that interval.*

Proof. Assume that f is continuous on $[a, b]$. Since $f(a) - 1 < f(a) < f(a) + 1$, it follows from Theorem 3.9.1 that on some interval $[a, c_1]$ we have $f(a) - 1 < f(x) < f(a) + 1$, which means that f is bounded on $[a, c_1]$.

Denote by x_0 the least upper bound of all numbers $c \le b$ such that f is bounded on $[a, c]$. Evidently $a < x_0 \le b$. We want to show that $x_0 = b$. Suppose $x_0 < b$. Because

$$f(x_0) - 1 < f(x_0) < f(x_0) + 1,$$

there exists some interval $[x_0, x_1]$ such that $f(x_0) - 1 < f(x) < f(x_0) + 1$ for every x in $[x_0, x_1]$. Thus

$$|f(x)| < |f(x_0)| + 1$$

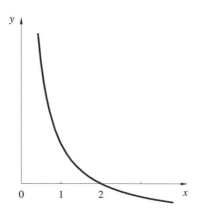

Figure 3.18 Graph of $y = \frac{2}{x} - 1$.

for all x in $[x_0, x_1]$. Since f is bounded on $[a, x_0]$, there exists a constant M such that

$$|f(x)| < M$$

for all x in $[a, x_1]$. But this contradicts the assumption that x_0 is the least upper bound of all numbers c such that f is bounded on $[a, c]$. This shows that $x_0 = b$, which implies that f is bounded on $[a, b]$.

Exercises 3.10

1. Give an example of an unbounded function on a bounded interval $[a, b]$.

2. True or false? If f is bounded on every closed subinterval of (a, b), then f is bounded on (a, b).

3. Let $\lim_{x \to x_0} f(x) = 0$ and let g be a bounded function in some interval $[x_0 - \delta, x_0 + \delta]$. Show that $\lim_{x \to x_0} f(x)g(x) = 0$.

4. Prove that if f and g are bounded on $[a, b]$, then $f + g$ and fg are bounded on $[a, b]$.

5. Decide whether the following are possible:

    (a) f is bounded, g is unbounded, and $f + g$ is bounded;

    (b) f and g are unbounded, but $f + g$ is bounded;

    (c) f is bounded, g is unbounded, and fg is bounded;

    (d) f and g are unbounded, but fg is bounded.

6. True or false? If f and g are bounded on $[a, b]$ and $g(x) \neq 0$ for all $x \in [a, b]$, then f/g is bounded on $[a, b]$.

7. True or false? If f is bounded on $[a, b]$, g is continuous on $[a, b]$ and $g(x) \neq 0$ for all $x \in [a, b]$, then f/g is bounded on $[a, b]$.

8. True or false? If f and g are bounded on (a, b), g is continuous on (a, b), and $g(x) \neq 0$ for all $x \in (a, b)$, then f/g is bounded on (a, b).

3.11 MINIMUM AND MAXIMUM OF A FUNCTION

The least value of the function $f(x) = x^2 - 1$ on the interval $[0, 2]$ is -1 and the greatest value is 3. On the open interval $(0, 2)$ the function assumes all the values in the interval $(-1, 3)$ and only those values. Thus on the open interval $(0, 2)$ the defined function does not have a greatest or least value.

The greatest value of a function is called its *maximum* and the least value its *minimum*. Each of them can be called an *extremum*. Existence of a minimum and a maximum on a closed bounded interval is one of the fundamental properties of continuous functions.

3.11.1 Theorem   *A function continuous on a closed bounded interval has a minimum and a maximum on that interval.*

Proof. Let W be the set of all values of a continuous function f on an interval $[a, b]$. By 3.10.2, W is bounded. Let K be the least upper bound of W. We want to show that there is x_0 in $[a, b]$ such that $f(x_0) = K$. Suppose, on the contrary, that $f(x) < K$ for all x in $[a, b]$. Then $K - f(x) > 0$ on $[a, b]$ and, consequently,

$$g(x) = \frac{1}{K - f(x)} > 0$$

on $[a, b]$. The function $g(x)$ is continuous on $[a, b]$, and, in view of 3.10.2, there is a positive number M such that $g(x) < M$ for every x in $[a, b]$; that is,

$$\frac{1}{K - f(x)} < M,$$

or equivalently

$$f(x) < K - \frac{1}{M}$$

for every x in $[a, b]$. But this contradicts the assumption that K is the least upper bound. Therefore, we must have $f(x_0) = K$ for some x_0 in $[a, b]$, which means that K is the maximum of f on $[a, b]$.

To prove that f has also a minimum it is enough to consider the function $-f$.

Exercises 3.11

1. Complete the proof of Theorem 3.11.1 by proving that a function continuous on a closed bounded interval has a minimum on that interval.

2. Give an example of a function on a closed bounded interval that does not have a maximum on that interval.

3. Give an example of a continuous function on an unbounded interval that does not have a maximum on that interval.

4. Give an example of a continuous function on a bounded interval that does not have a maximum on that interval.

5. True or false? If M is the maximum value of f, then $|M|$ is the maximum value of $|f|$.

6. True or false? If M is the maximum value of f and λ is a constant, then λM is the maximum value of λf.

7. True or false? If M_1 and M_2 are the maximum values of f_1 and f_2, respectively, then $M_1 + M_2$ is the maximum value of $f_1 + f_2$.

8. True or false? If M_1 and M_2 are the maximum values of f_1 and f_2, respectively, then $M_1 M_2$ is the maximum value of $f_1 f_2$.

3.12 COMPOSITE FUNCTIONS

If two functions $y = f(t)$ and $t = g(x)$ are given, we can define a new function $y = f(g(x))$, which is called a *composite function*. The composition of f and g is often denoted by $f \circ g$.

For example, if $f(t) = |t|$ and $g(x) = x^2 - x$, then $f(g(x)) = |x^2 - x|$. If $f(t) = t^2 - t$ and $g(x) = |x|$, then $f(g(x)) = x^2 - |x|$, since $|x|^2 = x^2$. The use of variables t and x is unessential. If $f(x) = \sqrt{x}$ and $g(x) = x^2 + 1$, then $f(g(x)) = \sqrt{x^2 + 1}$ and $g(f(x)) = x + 1$. Notice that in this example $f(g(x))$ is defined for all values of x, while $g(f(x))$ is defined only for non-negative values of x. When defining a composition of functions, we always have to be careful with the domain and the range of the composed functions. For instance, if $f(x) = \sqrt{x}$ and $g(x) = -x^2 - 1$, then the composition $f(g(x))$ is not defined for any $x \in \mathbb{R}$.

3.12.1. Theorem *If f is a function continuous at t_0 and $\lim_{x \to a} g(x) = t_0$, then*

(1)
$$\lim_{x \to a} f(g(x)) = f(t_0).$$

The assertion also holds if we replace $x \to a$ by $x \to a+$ or $x \to a-$.

Proof. Let ε be an arbitrary positive number. From the continuity of f it follows that there exists a number $\delta > 0$ such that $|f(t) - f(t_0)| < \varepsilon$ whenever $|t - t_0| < \delta$. Moreover, since $\lim_{x \to a+} g(x) = t_0$, there exists a number $\gamma > 0$ such that $|g(x) - t_0| < \delta$ whenever $a < x < a + \gamma$. Consequently, for every $\varepsilon > 0$ there exists $\gamma > 0$ such that $|f(g(x)) - f(t_0)| < \varepsilon$ whenever $a < x < a + \gamma$. This proves that $\lim_{x \to a+} f(g(x)) = f(t_0)$.

The proof for the left-hand limit is similar. Combining both results we obtain the assertion for the two-sided limit.

Equality (1) can be also written as

$$\lim_{x \to a} f(g(x)) = f(\lim_{x \to a} g(x)),$$

and similarly for the right-hand and the left-hand limits. This equation says that the evaluation of a limit and the evaluation of a function can be interchanged if the function is continuous. This enables us to calculate limits of composite functions. For example,

$$\lim_{x \to a} |x^2 - b| = |\lim_{x \to a} (x^2 - b)| = |a^2 - b|,$$

because the absolute value is a continuous function. On the other hand, for the function $[t]$ a similar calculation would be false, because

$$\left[\lim_{x \to 0} (-x^2)\right] = [0] = 0$$

and

$$\lim_{x \to 0} [-x^2] = -1,$$

because $\left[-x^2\right] = -1$ for $0 < x < 1$ and for $-1 < x < 0$. Changing the order in which the operations are performed leads to different results because the function $f(x) = [x]$ is not continuous at 0.

3.12.2 Theorem   *If g is continuous at a, and f is continuous at $t_0 = g(a)$, then the composite function $\phi(x) = f(g(x))$ is continuous at a.*

In short, but less precisely: *the composition of two continuous functions is continuous.*

Proof.   In view of 3.12.1 we have $\lim_{x \to a} f(g(x)) = f(g(a))$, which means precisely that ϕ is continuous at a.

3.12.3 Corollary   *A function f is continuous at a point a if and only if*

$$\lim_{x \to 0} f(a + x) = f(a).$$

Proof.   The function $g(x) = a + x$ is continuous everywhere. Thus, from the continuity of f at a, the continuity of the composite function $\phi(x) = f(a + x)$ follows. Conversely, the function f can be considered a composition of functions: $f(x) = \phi(x - a)$. Therefore, from the continuity of ϕ at 0 the continuity of f at a follows.

Exercises 3.12

1. An example in this section shows that $f \circ g$ need not equal $g \circ f$. How about the associativity of composition: $f \circ (g \circ h) = (f \circ g) \circ h$?
2. Does there exist a function f such that $f \circ g = g \circ f = g$ for any function g?
3. Decide whether the following are possible:
   (a) f is continuous and g is not continuous, but $f \circ g$ is continuous.
   (b) f is continuous and g is not continuous, but $g \circ f$ is continuous.
   (c) Neither f nor g are continuous, but $f \circ g$ is continuous.

3.13 ASCENT AND SPEED

The road drawn in Fig. 3.19 ascends 1 meter for each 12 meters passed horizontally. We say that the slope of the road is $\frac{1}{12}$. In order to find the slope we do not have to take a 12-meter-long segment of the road. For instance, noticing that for every 6 meters in the horizontal direction the road rises $\frac{1}{2}$ meter, then the slope can be found by dividing $\frac{1}{2}$ by 6 which again gives $\frac{1}{12}$. The slope does not depend on the segment of road chosen. Note that also the use of meters to measure length is unessential. If feet were used we would get exactly the same result.

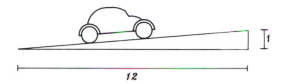

Figure 3.19 Road with a constant slope.

The situation changes if the inclination of the road is different in different places, as in Fig. 3.20. The average slope for the 12-meter-segment marked on the picture can be found by dividing 4 by 12, giving $\frac{1}{3}$. On the other hand, if we consider only the second half of the segment, the average slope is $\frac{1}{6}$.

It is clear from the picture that the slope is greatest at the beginning of the road and then it decreases gradually. Intuitively, the value for the slope can be assigned to every point of the road. How can it be expressed numerically?

Assume that the shape of the road in Fig. 3.20 is represented approximately by the function

$$f(x) = \frac{2}{3}x - \frac{1}{36}x^2.$$

We wish to find the slope at $x = 6$. The elevation at this point is 3. If we move to the right a distance h, then we shall be lifted to the height

$$f(6 + h) = \frac{2}{3}(6 + h) - \frac{1}{36}(6 + h)^2 = 3 + \frac{1}{3}h - \frac{1}{36}h^2.$$

The difference in elevation is then

$$f(6 + h) - f(6) = \frac{1}{3}h - \frac{1}{36}h^2,$$

and the average slope on the interval $[6, 6 + h]$ is

$$\frac{f(6 + h) - f(6)}{h} = \frac{1}{3} - \frac{1}{36}h.$$

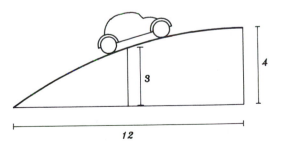

Figure 3.20 Road with a varying slope.

If h tends to 0, we obtain

$$\lim_{h \to 0} \frac{f(6 + h) - f(6)}{h} = \frac{1}{3}.$$

The number $\frac{1}{3}$ is the slope at $x = 6$.

The slope at a point can be also interpreted as the slope of the tangent line at that point. In Fig. 3.21 the straight line PQ has the slope

$$\frac{f(x + h) - f(x)}{h}.$$

If point P is fixed and Q approaches P, then the secant line PQ gradually "approaches" the tangent line at P. The slope of the tangent line can be found by calculating the limit

$$\lim_{h \to 0} \frac{f(x + h) - f(x)}{h}.$$

Suppose a train leaves a railway station. Let us denote its distance from that station by y. The distance y depends on time t, so we can write $y = f(t)$. Let t_0 be a fixed instant. Between instances t_0 and $t_0 + h$ the average speed of the train is

$$\frac{f(t_0 + h) - f(t_0)}{h}.$$

The limit

$$\lim_{h \to 0} \frac{f(t_0 + h) - f(t_0)}{h}$$

represents the instantaneous speed of the train at the instant t_0.

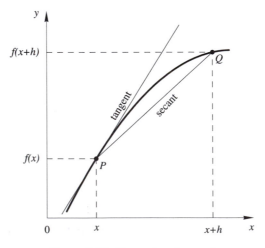

Figure 3.21 Tangent and secant.

Exercises 3.13 ───────────────────────────────

1. Let $f(x) = x^2 + 1$. Find the slope of the tangent to the graph of f at $x = 2$.
2. Let $f(x) = (x + 1)^2$. Find the slope of the tangent to the graph of f at $x = 2$.

3.14 DERIVATIVES

3.14.1 Definition   If the limit

$$(1) \qquad \lim_{h \to 0} \frac{f(x + h) - f(x)}{h}$$

exists it is called the *derivative* of f at x. If the derivative of f exists at a point x, we say that f is *differentiable* at x. If f is differentiable at every point of an interval, we simply say that f is differentiable on that interval.

Because the limit (1) generally has different values at different points, it is a function of x. This function will be denoted by f'.

3.14.2 Examples

1. If $f(x) = 1/x$, then

$$f'(x) = \lim_{h \to 0} \frac{\dfrac{1}{x + h} - \dfrac{1}{x}}{h} = -\frac{1}{x^2}.$$

In this example the function is differentiable at every point of its domain, or in other words for $x \neq 0$.

2. If $f(x) = x$, then

$$f'(x) = \lim_{h \to 0} \frac{x + h - x}{h} = 1.$$

3. If f is a constant function $f(x) = c$, then

$$f'(x) = \lim_{h \to 0} \frac{c - c}{h} = 0. \qquad \blacksquare$$

Examples 2 and 3 are particularly important, for they are the base for the calculation of many other, more complicated, derivatives. For this reason it is advisable to remember them well.

Exercises 3.14 ───────────────────────────────

1. Using the definition find the derivatives of the following functions:

$$f(x) = 2x; \qquad f(x) = x^2; \qquad f(x) = \frac{1}{x^2}.$$

2. Find the equation of the tangent line to the graph of $y = x^3$ at $x = 1$.

3. Find the point on the parabola $y = x^2$ that is closest to the line $y = x - 1$. (*Hint:* Use the fact that the tangent line at that point is parallel to the line $y = x - 1$.)

4. Let f be a differentiable function such that $f(x) \neq 0$ for all $x \in \mathbb{R}$. Use the definition of the derivative to prove that the function $g(x) = 1/f(x)$ is differentiable, and express g' in terms of f and f'.

5. Let f be a differentiable function. Use the definition of the derivative to prove that the function $g(x) = (f(x))^2$ is differentiable, and express g' in terms of f and f'.

6. Give an example of a function that is not differentiable at a point.

3.15 DIFFERENTIABILITY AND CONTINUITY

3.15.1 Theorem   *A function differentiable at a point is continuous at that point.*

Proof.   For every $h \neq 0$, we have

$$f(x + h) = \frac{f(x + h) - f(x)}{h} h + f(x).$$

If f is differentiable at x, then

$$\lim_{h \to 0} f(x + h) = \lim_{h \to 0} \frac{f(x + h) - f(x)}{h} \cdot \lim_{h \to 0} h + \lim_{h \to 0} f(x)$$

$$= f'(x) \cdot 0 + f(x) = f(x),$$

proving, by 3.12.3, continuity at x.

The converse is not true. For instance, the function $f(x) = |x|$ (see Fig. 3.22) is continuous at 0 (it is continuous everywhere) but it is not differentiable at 0. Indeed,

$$\lim_{h \to 0+} \frac{|0 + h| - 0}{h} = 1 \quad \text{and} \quad \lim_{h \to 0-} \frac{|0 + h| - 0}{h} = -1,$$

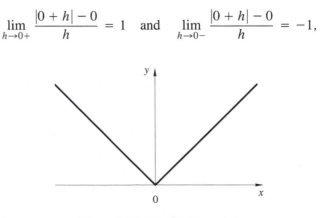

Figure 3.22 Graph of $y = |x|$.

and, therefore, the limit does not exist. This has also a geometrical meaning: the graph of $f(x) = |x|$ has an edge at 0, so there is no tangent line at that point. Karl Weierstrass (1815–1897) constructed a function that is continuous everywhere but does not have a derivative at any point.

Exercises 3.15

1. Give an example of a function that is not differentiable at integral points $x = 0$, $\pm 1, \pm 2, \ldots$, but is differentiable at every other point.

2. Let
$$f(x) = \begin{cases} x^2 & \text{if } x \text{ is a rational number,} \\ 0 & \text{if } x \text{ is an irrational number.} \end{cases}$$
Show that f is differentiable at $x = 0$ but not differentiable at any other point.

3. Possible or not? f is not differentiable at 0 but f^2 is.

4. Give an example of a function differentiable everywhere except at two given points a and b.

3.16 DERIVATIVE OF A PRODUCT

To differentiate the product of two functions we use the following rule:

3.16.1 Theorem *The product fg of two functions f and g differentiable at a point is differentiable at that point and its derivative is given by*

(1)
$$(fg)' = f'g + fg'.$$

Proof. Let $\phi(x) = f(x)g(x)$. Then, for $h \neq 0$,

$$\frac{\phi(x + h) - \phi(x)}{h}$$

$$= \frac{f(x + h)g(x + h) - f(x)g(x)}{h}$$

$$= \frac{f(x + h)g(x + h) - f(x)g(x + h) + f(x)g(x + h) - f(x)g(x)}{h}$$

$$= \frac{f(x + h) - f(x)}{h}g(x + h) + f(x)\frac{g(x + h) - g(x)}{h}.$$

Since g is differentiable, and thus continuous, we have $\lim_{h \to 0} g(x + h) = g(x)$. Hence

$$\lim_{h \to 0} \frac{\phi(x + h) - \phi(x)}{h} = \lim_{h \to 0} \frac{f(x + h) - f(x)}{h} \lim_{h \to 0} g(x + h)$$

$$+ \lim_{h \to 0} f(x) \lim_{h \to 0} \frac{g(x + h) - g(x)}{h}$$

$$= f'(x)g(x) + f(x)g'(x).$$

Applying (1) successively we get

$$(x^2)' = (xx)' = x + x = 2x;$$
$$(x^3)' = (x^2 x)' = 2xx + x^2 = 3x^2.$$

In general, for every natural number n,

(2) $$(x^n)' = nx^{n-1}.$$

In fact, this formula holds for $n = 1$, since $x^1 = x$ and $x^0 = 1$. Assume now that (2) holds for some $n \in \mathbb{N}$. Then

$$\left(x^{n+1}\right)' = (x^n x)' = (x^n)'x + x^n = (n+1)x^n.$$

Hence, by induction, formula (2) follows for all natural numbers n.

If $f(x) = c$ is a constant function, then $f'(x) = (c)' = 0$ and hence, by (1), we obtain $(f(x)g(x))' = (cg(x))' = (c)'g(x) + cg'(x) = cg'(x)$. We thus have the following theorem:

3.16.2 Theorem $(cg)' = cg'$ *(where c is a constant).*

Exercises 3.16

1. Prove 3.16.2 using the definition of the derivative.

2. Let f be a differentiable function such that $f(x) \neq 0$ for all $x \in \mathbb{R}$. Use Theorem 3.16.1 to prove that

    $$\left(\frac{1}{f}\right)' = -\frac{f'}{f^2}.$$

    (Hint: $1 = f \cdot \frac{1}{f}$.)

3. Use the induction principle and Theorem 3.16.1 to prove that if f is differentiable then, for any $n \in \mathbb{N}$, $g(x) = (f(x))^n$ is differentiable and $g'(x) = n(f(x))^{n-1}f'(x)$.

4. Possible or not? f is differentiable at a, g is not differentiable at a, but the product fg is differentiable at a.

5. Find a formula for the derivative of the product of three functions.

3.17 DERIVATIVE OF A SUM AND A DIFFERENCE

3.17.1 Theorem   *The sum $f + g$ of two functions f and g differentiable at a point is differentiable at that point and its derivative is given by*

$$(f + g)' = f' + g'.$$

Proof. Let $\phi(x) = f(x) + g(x)$. Then

$$\frac{\phi(x+h) - \phi(x)}{h} = \frac{f(x+h) - f(x)}{h} + \frac{g(x+h) - g(x)}{h}.$$

The assertion follows by passing to the limit as $h \to 0$.

3.17.2 Theorem   *The difference $f - g$ of two functions f and g differentiable at a point is differentiable at that point and its derivative is given by*

$$(f - g)' = f' - g'.$$

Proof.   Using 3.16.2 and 3.17.1 we have

$$(f - g)' = (f + (-1)g)' = f' + (-1)g' = f' - g'.$$

Using the rules of this and the preceding section we can easily calculate derivatives of more complicated functions, such as

$$(3x^2 - x + 2)' = 6x - 1;$$

$$(6x^5 - \frac{x^4}{4} - \frac{x}{2} - 2)' = 30x^4 - x^3 - \frac{1}{2}.$$

Exercises 3.17

1.   Prove that polynomial functions are differentiable everywhere.

2.   Possible or not? f and g are not differentiable at a, but $f + g$ is differentiable at a.

3.18 DERIVATIVE OF A QUOTIENT

3.18.1 Theorem   *If functions f and g are differentiable at a point and g is different from zero at that point, then the quotient f/g is also differentiable at that point and*

(1) $$\left(\frac{f}{g}\right)' = \frac{f'g - fg'}{g^2}.$$

The proof follows from the following equality

$$\frac{\frac{f(x+h)}{g(x+h)} - \frac{f(x)}{g(x)}}{h} = \frac{\frac{f(x+h) - f(x)}{h} g(x) - f(x) \frac{g(x+h) - g(x)}{h}}{g(x)g(x+h)}$$

by letting $h \to 0$ (that is, by taking the limit as $h \to 0$).

3.18.2 Examples

1. $\left(\dfrac{x^2 + 1}{x^2 - x + 1}\right)' = \dfrac{2x(x^2 - x + 1) - (x^2 + 1)(2x - 1)}{(x^2 - x + 1)^2} = \dfrac{1 - x^2}{(x^2 - x + 1)^2}.$

2. $\left(\dfrac{1}{x^2 + 1}\right)' = \dfrac{0 \cdot (x^2 + 1) - 1 \cdot 2x}{(x^2 + 1)^2} = \dfrac{-2x}{(x^2 + 1)^2}.$

If the numerator is a constant, it is more convenient to use the following simplified formula:

3.18.3 *Let c be a constant. Then*

$$\left(\frac{c}{g}\right)' = -\frac{cg'}{g^2}.$$

3.18.3 can be used to prove the formula for the derivative of negative integral powers of x. Indeed,

$$\left(\frac{1}{x^n}\right)' = \frac{-nx^{n-1}}{(x^n)^2} = -\frac{n}{x^{n+1}}.$$

—Why do we only use rational functions in all our examples? If we admitted other functions like sine or logarithm we could greatly extend the stock of examples and make it more interesting.

—What is the sine?

—The ratio of the side opposite the acute angle of a right triangle to its hypotenuse.

—What is a hypotenuse?

—A lecture on geometry would be needed to explain all this.

—Precisely. We do not admit any geometrical definitions in this book. Definitions of sine, logarithm, and other so-called elementary functions will be supplied later and based on concepts of analysis.

Exercises 3.18

1. Use 3.16.1 to prove 3.18.1.

2. Show that a rational function is differentiable at every point of its domain.

3. Show that the derivative of a rational function is a rational function whose domain is the same as that of the original function.

3.19 MINIMUM AND MAXIMUM OF A FUNCTION (CONTINUED)

In Section 3.11 we proved the existence of the minimum and the maximum of a continuous function on a closed interval. The following theorem can be used to find their values.

3.19.1 Theorem   *If a function f defined on an interval (a, b) has an extremum at $x_0 \in (a, b)$ and is differentiable at x_0, then $f'(x_0) = 0$.*

Proof.   Suppose f has a minimum at x_0; then for every $x_0 + h$ in the considered interval we have $f(x_0 + h) - f(x_0) \geq 0$, and consequently

$$\frac{f(x_0 + h) - f(x_0)}{h} \leq 0 \quad \text{if } h < 0,$$

$$\frac{f(x_0 + h) - f(x_0)}{h} \geq 0 \quad \text{if } h > 0.$$

By letting $h \to 0$ we get, in view of 3.6.1,

$$f'(x_0) \leq 0 \quad \text{and} \quad f'(x_0) \geq 0,$$

which proves that $f'(x_0) = 0$.

If f has a maximum at x_0, this proof requires only some obvious modifications.

From 3.19.1 it follows that a function defined on a closed interval may have its maximum and minimum only

(a)   where the derivative equals 0;

(b)   where the derivative does not exist; or

(c)   at the ends of the interval.

The function cannot have an extremum at any other point.

3.19.2 Example   Find the minimum and the maximum of the function $f(x) = x^2 - 2x - 3$ on the interval $[-2, 2]$. (See Fig. 3.23.)

The function is differentiable everywhere: $f'(x) = 2x - 2$. The only point at which $f'(x) = 0$ is $x = 1$. Thus f can have extrema at $x = 1$ or at the endpoints $x = -2$ and $x = 2$. Since $f(1) = -4$, $f(-2) = 5$, and $f(2) = -3$, we infer that the minimum is -4 and the maximum is 5. ∎

3.19.3 Example   Find the minimum and the maximum of $f(x) = |2x + 1|$ on the interval $[-1, 1]$. (See Fig. 3.24.)

Since

$$f(x) = \begin{cases} 2x + 1 & \text{for } x \geq -\frac{1}{2}, \\ -2x - 1 & \text{for } x < -\frac{1}{2}, \end{cases}$$

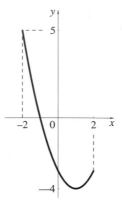

Figure 3.23 Graph
of $y = x^2 - 2x - 3$
for $x \in [-2, 2]$.

we have

$$f'(x) = \begin{cases} 2 & \text{for } x > -\frac{1}{2}, \\ -2 & \text{for } x < -\frac{1}{2}, \end{cases}$$

and at $x = -\frac{1}{2}$ there is no derivative. Since the derivative is never equal to 0, the extrema can be reached only at points $x = -\frac{1}{2}$, $x = -1$, or $x = 1$. Since $f(-\frac{1}{2}) = 0$, $f(-1) = 1$, and $f(1) = 3$, the minimum is 0 and the maximum is 3.

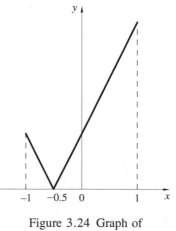

Figure 3.24 Graph of
$y = |2x - 1|$ for $x \in [-1, 1]$.

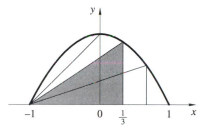

Figure 3.25 The triangle with the
maximal area in Example 3.19.4

3.19.4 Example   Refer to Fig. 3.25. In the parabola $y = 1 - x^2$, there is
inscribed a right triangle so that one of its sides is perpendicular to the x-axis. Find
the triangle so inscribed whose area is maximal.

It is evident that one of the vertices must lie at the point $(1, 0)$ or $(-1, 0)$. Because
of the symmetry we can assume the second case. Then the point $(-1, 0)$ is also the
end-point of the hypotenuse. Let the other end of the hypotenuse have coordinates
(x, y). Because this end lies on the arc of the parabola, we have $y = 1 - x^2$.

The area of the triangle is $\frac{1}{2}(1 + x)y$. We have to find the maximum of the
function

$$P(x) = \frac{1}{2}(1 + x)(1 - x^2)$$

on the interval $(-1, 1)$. We have

$$P'(x) = \frac{-1}{2}(3x - 1)(x + 1).$$

Therefore, $P'(x) = 0$ for $x = \frac{1}{3}$ and $x = -1$. Looking for the extrema we have to
take into account points -1, $\frac{1}{3}$, and 1. Since $P(-1) = P(1) = 0$ and $P(\frac{1}{3}) = \frac{16}{27}$,
we infer that the maximal area is $\frac{16}{27}$ and is reached at $x = \frac{1}{3}$.   ■

Exercises 3.19

1. Prove 3.19.1 for the maximum.
2. Show that $x + 1/x \geq 2$ for all $x > 0$.
3. Find p and q such that the function $f(x) = x^2 + px + q$ assumes its minimum
    value, 3, at $x = 1$.

3.20 MEAN VALUE THEOREM

Consider an arc of a curve and the chord joining its ends (see Fig. 3.26). If at each
point of the arc there is a tangent, then from among those tangents we can select
one that is parallel to the chord. This theorem, which is obvious geometrically, will
be formulated and proved analytically.

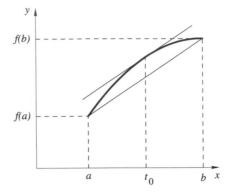

Figure 3.26 Illustration of the mean
value theorem.

3.20.1 Theorem (mean value theorem)   *If a function f is continuous on* $[a, b]$
and differentiable on (a, b)*, then there exists at least one point* $t_0 \in (a, b)$ *such that*

(1) $$\frac{f(b) - f(a)}{b - a} = f'(t_0).$$

Proof.   We introduce an auxiliary function:

$$\phi(x) = f(x) - \frac{f(b) - f(a)}{b - a}(x - a).$$

This function is also continuous on $[a, b]$ and differentiable on (a, b). Moreover,
$\phi(a) = \phi(b)$.

Let M denote the maximum and m the minimum of ϕ on $[a, b]$. If $M = m$,
then ϕ is a constant function and $\phi(x) = \phi(a)$ for every $x \in [a, b]$, and hence
$\phi'(x) = 0$ everywhere on $[a, b]$. If $M > m$, then either $M > \phi(a)$ or $m < \phi(a)$.
In the first case the function ϕ reaches the maximum at some point t_0 inside the
interval, and by 3.19.1, $\phi'(t_0) = 0$. Similarly, if $m < \phi(a)$, the function ϕ has a
minimum at a point $t_0 \in (a, b)$, and thus $\phi'(t_0) = 0$. Thus, in any case there exists
a number $t_0 \in (a, b)$ such that $\phi'(t_0) = 0$. But

$$\phi'(x) = f'(x) - \frac{f(b) - f(a)}{b - a},$$

and hence

$$0 = f'(t_0) - \frac{f(b) - f(a)}{b - a},$$

proving the theorem.

The mean value theorem is often stated in the following form:

3.20.2 Theorem   *If a function f is continuous on* $[a, b]$ *and differentiable on*
(a, b)*, and* $x, x_0 \in [a, b]$*, then*

$$f(x) - f(x_0) = (x - x_0)f'(t_0) \quad \textit{for some } t_0 \textit{ between } x \textit{ and } x_0.$$

This form is often used if x is a variable in a fixed interval. One has to remember that in 3.20.2 the value t_0 depends on x. Overlooking this fact may lead to false results; this has happened even to prominent authors.

3.20.3 Theorem   *If f is a function continuous on an interval $[a, b]$ and $f'(x) = 0$ for each $x \in (a, b)$, then f is a constant function on $[a, b]$.*

Proof.   If $a \le x_1 < x_2 \le b$, then there exists a point t_0 in (x_1, x_2) such that

$$\frac{f(x_2) - f(x_1)}{x_2 - x_1} = f'(t_0) = 0.$$

Hence $f(x_1) = f(x_2)$.

3.20.4 Theorem   *If two functions are continuous on an interval $[a, b]$ and have the same derivative on (a, b), then they differ by an additive constant.*

Proof.   The difference of the functions satisfies 3.20.3 and thus it is a constant.

Exercises 3.20

1. Show that if f is not differentiable at one point of (a, b) then (1) need not hold.

2. Give an example of a nonconstant function f such that $f'(x) = 0$ for all points in its domain.

3. Let f be a differentiable function in an interval $(-\alpha, \alpha)$. Prove that if $f(0) = 0$ and $|f'(x)| \le \beta$ for all $x \in (-\alpha, \alpha)$, then $|f(x)| \le \alpha\beta$ for all $x \in (-\alpha, \alpha)$.

4. Prove that if $f(0) = 0$ and $|f'(x)| \le M$ for all $x \in \mathbb{R}$, then $|f(x)| \le M|x|$ for all $x \in \mathbb{R}$.

5. Let f and g be differentiable on $(0, \infty)$. Prove that if $f(0) = g(0)$ and $f'(x) \le g'(x)$ for all $x > 0$ then $f(x) \le g(x)$ for all $x > 0$.

6. Prove Cauchy's generalized mean value theorem: If the functions f and g are continuous on $[a, b]$ and differentiable on (a, b), then there exists $c \in (a, b)$ such that

$$(f(b) - f(a))\,g'(c) = (g(b) - g(a))\,f'(c).$$

    (*Hint:* Consider $\phi(x) = (f(b) - f(a))\,g(x) - (g(b) - g(a))\,f(x)$.)

3.21 INCREASING AND DECREASING FUNCTIONS

A function is called *increasing* (see Fig. 3.27) if its value increases whenever x increases. A function is called *decreasing* (see Fig. 3.28) if its value decreases whenever x increases. More precisely:

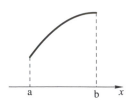

Figure 3.27 Function
increasing on (a, b).

3.21.1. Definition   A function f is *increasing* on an interval if for any two points $x_1 < x_2$ of that interval $f(x_1) < f(x_2)$. A function f is *decreasing* on an interval if for any two points $x_1 < x_2$ of that interval $f(x_1) > f(x_2)$.

From the above definition it follows that an increasing function or a decreasing function in an interval can assume each value only once.

Functions can be increasing on some intervals and decreasing on other intervals. For instance, the function $f(x) = x^2$ is increasing on $(0, \infty)$ and decreasing on $(-\infty, 0)$.

3.21.2 Theorem   *If the derivative of f is positive on an interval, then f is increasing on that interval. If the derivative of f is negative on an interval, then f is decreasing on that interval.*

Proof.   Assume that $f'(x) > 0$ for every x in some interval. Suppose that f is not increasing in that interval; that is, there are $x_1 < x_2$ such that $f(x_1) \geq f(x_2)$. Then

$$\frac{f(x_2) - f(x_1)}{x_2 - x_1} \leq 0.$$

By the mean value theorem there is $t_0 \in (x_1, x_2)$ such that

$$f'(t_0) = \frac{f(x_2) - f(x_1)}{x_2 - x_1},$$

which contradicts the assumption that $f'(x) > 0$. Thus the function is increasing.

The proof of the second part is similar.

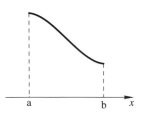

Figure 3.28 Function
decreasing on (a,b).

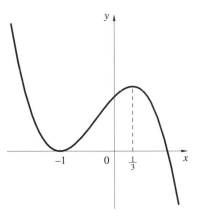

Figure 3.29 Graph of
$y = (1 + x)(1 - x^2)$.

The above theorem can be used to investigate behavior of functions. For example the function $f(x) = (1 + x)(1 - x^2)$ (see Fig. 3.29) has derivative $f'(x) = -(3x - 1)(x + 1)$, which is positive on $(-1, \frac{1}{3})$ and negative on $(-\infty, -1)$ and on $(\frac{1}{3}, \infty)$. Hence f is increasing on $(-1, \frac{1}{3})$ and decreasing on $(-\infty, -1)$ and $(\frac{1}{3}, \infty)$.

In Section 3.9 we proved that the equation

(1)
$$x^5 + x - 1 = 0$$

has a solution in $(0, 1)$. The function $f(x) = x^5 + x - 1$ has derivative $f'(x) = 5x^4 + 1$. Because this derivative is positive everywhere, the function f is increasing. Thus the solution of (1) is unique.

The converse of Theorem 3.21.2 is false. For instance, the function $f(x) = x^3$ is increasing everywhere, although its derivative $f'(x) = 3x^2$ equals 0 at 0 (see Fig. 3.30).

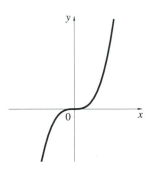

Figure 3.30 Graph of
$y = x^3$.

Exercises 3.21

1. Complete the proof of 3.21.2.

2. Let f be continuous on $[0, \infty)$ and differentiable on $(0, \infty)$. Show that if $f(0) = 0$ and f' is increasing on $(0, \infty)$, then the function $g(x) = f(x)/x$ is increasing on $(0, \infty)$.

3. True or false? If f is an increasing function, then f' is an increasing function.

3.22 NONINCREASING AND NONDECREASING FUNCTIONS

A function is called *nonincreasing* or *nondecreasing* if its value does not increase or does not decrease, respectively, when x increases. More precisely:

3.22.1 Definition   A function f is called *nonincreasing* on an interval if for any pair of points $x_1 < x_2$ in that interval we have $f(x_1) \geq f(x_2)$. A function f is called *nondecreasing* in an interval if for any pair of points $x_1 < x_2$ in that interval $f(x_1) \leq f(x_2)$. If a function is nonincreasing or nondecreasing, we say it is *monotone*.

Every increasing function is nondecreasing, but not conversely. Similarly, every decreasing function is nonincreasing. If f is increasing, $-f$ is decreasing. If f is nonincreasing, $-f$ is nondecreasing. Also, if f is decreasing or nondecreasing, $-f$ is increasing or nonincreasing, respectively.

3.22.2 Theorem   *If $f'(x) \geq 0$ in some interval, then f is nondecreasing on that interval. If $f'(x) \leq 0$ on some interval, then f is nonincreasing on that interval.*

Proof.   Suppose that $f'(x) \geq 0$ on an interval and $f(x_1) > f(x_2)$ for some $x_1 < x_2$ in that interval. Then, for some $x_1 < t_0 < x_2$, we have

$$f'(t_0) = \frac{f(x_2) - f(x_1)}{x_2 - x_1} < 0,$$

contrary to the assumption. The proof of the second part is similar.

In contrast to Theorem 3.21.2, Theorem 3.22.2 can be reversed.

3.22.3 Theorem   *If f is differentiable and nonincreasing on an interval, then $f'(x) \leq 0$ on that interval. Similarly, if f is differentiable and nondecreasing on an interval, then $f'(x) \geq 0$ on that interval.*

Proof.   In the first case we have

$$\frac{f(x + h) - f(x)}{h} \leq 0$$

for x and $x + h$ in the interval ($h \neq 0$). Hence, letting $h \to 0$, we get $f'(x) \leq 0$. In the second case the proof is similar.

Exercises 3.22

1. Complete the proof of 3.22.2.
2. Complete the proof of 3.22.3.

3.23 LIMIT AT INFINITY

If in the function $f(x) = 1/x$ we substitute for x greater and greater positive numbers, then the values of the function become closer and closer to 0. We then say that the function has the limit 0 at infinity. More generally, there is the following definition:

3.23.1 Definition   We say that a function f tends to a limit k as $x \to \infty$, if for every $\varepsilon > 0$ there exists a number x_0 such that $|f(x) - k| < \varepsilon$ for $x > x_0$. We then write $\lim_{x\to\infty} f(x) = k$.

For example: $\lim_{x\to\infty} \dfrac{1}{x} = 0$.

3.23.2 Theorem $\lim_{x\to\infty} f(x) = k$ *if and only if* $\lim_{x\to 0+} f\left(\dfrac{1}{x}\right) = k$.

Proof.   If $\lim_{x\to\infty} f(x) = k$, then for every $\varepsilon > 0$ there exists a positive number x_0 such that

$$(1) \qquad\qquad |f(x) - k| < \varepsilon$$

for $x > x_0$. If $x = 1/t$ and $1/x_0 = \delta$, then for $0 < t < \delta$ we have $1/t > x_0$ and, by (1),

$$(2) \qquad\qquad |f(1/t) - k| < \varepsilon.$$

Thus, for every $\varepsilon > 0$ there exists $\delta > 0$ such that (2) holds for $0 < t < \delta$. This proves that $\lim_{x\to\infty} f(x) = k$ implies that $\lim_{x\to 0+} f(1/x) = k$.

If we assume, conversely, that $\lim_{x\to 0+} f(1/x) = k$, then for every $\varepsilon > 0$ there exists a number $\delta > 0$ such that (2) holds for $0 < t < \delta$. Again letting $x = 1/t$ and $1/x_0 = \delta$, the inequality $0 < t < \delta$ is equivalent to $x > x_0$ and (2) to (1). Hence, for every $\varepsilon > 0$ there exists a number x_0 such that (1) holds for $x > x_0$. Thus we have proved that $\lim_{x\to 0+} f(1/x) = k$ implies $\lim_{x\to\infty} f(x) = k$.

For example:

$$\lim_{x\to\infty} \frac{x^2 - 2x}{2x^2 - 3x - 1} = \lim_{x\to 0+} \frac{\dfrac{1}{x^2} - \dfrac{2}{x}}{\dfrac{2}{x^2} - \dfrac{3}{x-1}} = \lim_{x\to 0+} \frac{1 - 2x}{2 - 3x - x^2} = \frac{1}{2}.$$

The following important theorem is a consequence of the Dedekind Axiom of the real numbers. Note that f need not be a continuous function.

3.23.3 Theorem   *If a function f is monotone and bounded on an interval (a, ∞), then the limit $\lim_{x\to\infty} f(x)$ exists.*

Proof.   The boundedness of f means that the set Z of values that f assumes on (a, ∞) is bounded. Let K be the least upper bound of Z and k its greatest lower bound. For every $\varepsilon > 0$ there exists a number $x_0 > a$ such that $K - \varepsilon < f(x_0) \leq K$. If f is nondecreasing, then we also have $K - \varepsilon < f(x) \leq K$ for every $x > x_0$. Hence $|f(x) - K| < \varepsilon$ for $x > x_0$, proving that $\lim_{x \to \infty} f(x) = K$. If f is nonincreasing, then $\lim_{x \to \infty} f(x) = k$. Indeed, for every $\varepsilon > 0$ there exists a number x_0 such that $k \leq f(x_0) < k + \varepsilon$. Then $k \leq f(x) < k + \varepsilon$ for $x > x_0$, proving that $\lim_{x \to \infty} f(x) = k$.

We will also need the *limit at negative infinity,* which can be defined as follows:

$$\lim_{x \to -\infty} f(x) = \lim_{x \to \infty} f(-x).$$

This limit has properties similar to those of the limit at infinity.

3.23.4. $\lim_{x \to -\infty} f(x) = k$ *if and only if* $\lim_{x \to 0-} f\left(\dfrac{1}{x}\right) = k$.

3.23.5.   *For a monotone and bounded function f defined on an interval $(-\infty, a)$ the limit* $\lim_{x \to -\infty} f(x)$ *exists.*

The proofs can be obtained as corollaries of 2.23.2 and 2.23.3.

Exercises 3.23

1.   Prove 3.23.4.
2.   Prove 3.23.5.
3.   Discuss the limit

$$\lim_{x \to \infty} \frac{a_n x^n + \cdots + a_1 x + a_0}{b_m x^m + \cdots + b_1 x + b_0}.$$

4.   Discuss the limit

$$\lim_{x \to -\infty} \frac{a_n x^n + \cdots + a_1 x + a_0}{b_m x^m + \cdots + b_1 x + b_0}.$$

5.   Find all theorems in this chapter having proofs based on the Dedekind Axiom.

CHAPTER 4

SEQUENCES AND SERIES

4.1 SEQUENCES

We have already used sequences in Section 2.12. Formally, a *sequence* is a function defined on the set $\mathbb{N}$ of natural numbers. The value of that function at the point n is called the nth term of the sequence and is denoted by a_n. The letters a and n play the same role as f and x when denoting a function. We will write "a sequence $a_1, a_2, \ldots$" or "a sequence $\{a_n\}$." The braces are used to distinguish the sequence $\{a_n\}$ from its nth term a_n.

4.1.1 Definition   We say that a sequence $\{a_n\}$ has a *limit* a, if for every $\varepsilon > 0$ there is a number n_0 such that $|a_n - a| < \varepsilon$ for all $n > n_0$. We then write

$$\lim_{n \to \infty} a_n = a \quad \text{or} \quad a_n \to a.$$

Note that the definition is exactly the same as for function: one only replaces f by a and x by n. The symbol $a_\infty = a$ is not used.

One can also say that the sequence $\{a_n\}$ tends to a or $\{a_n\}$ *converges* to a. Sequences that have limits are called *convergent*. Sequences that are not convergent are called *divergent*.

A sequence whose terms are all equal is called a *constant sequence*. Clearly, the sequence $a, a, a, \ldots$ converges to a.

A sequence $\{a_n\}$ is called *bounded* if there exists a constant M such that $|a_n| < M$ for all $n \in \mathbb{N}$.

4.1.2 Theorem   *Every convergent sequence is bounded.*

Proof.   Let ε be an arbitrary positive fixed number. If a is the limit of $\{a_n\}$ then there exists a number n_0 such that $|a_n - a| < \varepsilon$ for $n > n_0$. Hence $|a_n| \leq |a| + \varepsilon$ for $n > n_0$. On the other hand, the set of numbers $|a_n|$ with indices $n \leq n_0$ is bounded by a number M_0. Denoting by M the greater of numbers $|a| + \varepsilon$ and M_0, we have $|a_n| < M$ for all $n \in \mathbb{N}$, which proves the boundedness of $\{a_n\}$.

4.1.3 Theorem   *If $|a_n| \leq b_n$ for every $n \in \mathbb{N}$ and $\lim_{n \to \infty} b_n = 0$ then $\lim_{n \to \infty} a_n = 0$.*

Proof. For every $\varepsilon > 0$ there is a number n_0 such that for $n \geq n_0$ we have $b_n = |b_n - 0| < \varepsilon$ and consequently $|a_n - 0| = |a_n| \leq b_n < \varepsilon$, which proves the theorem.

4.1.4 Theorem *If* $\lim_{n \to \infty} a_n = a$ *and* $\lim_{n \to \infty} b_n = b$, *then*

(I) $\displaystyle \lim_{n \to \infty} (a_n + b_n) = a + b$;

(II) $\displaystyle \lim_{n \to \infty} (a_n - b_n) = a - b$;

(III) $\displaystyle \lim_{n \to \infty} a_n b_n = ab$.

Moreover, for $b_n \neq 0$ *and* $b \neq 0$,

(IV) $\displaystyle \lim_{n \to \infty} \frac{a_n}{b_n} = \frac{a}{b}$.

Proof. This theorem is similar to Theorem 3.5.1 on limits of functions. If there is a function $f(x)$ such that $f(n) = a_n$ and the limit $\lim_{x \to \infty} f(x)$ exists then obviously the limit $\lim_{n \to \infty} a_n$ also exists and the limits are equal. To prove Theorem 4.1.4 assume that the limits $\lim_{n \to \infty} a_n = a$ and $\lim_{n \to \infty} b_n = b$ exist and that there are functions $f(x)$ and $g(x)$ such that $f(n) = a_n$, $g(n) = b_n$, $\lim_{x \to \infty} f(x) = a$, and $\lim_{x \to \infty} g(x) = b$. Then by Theorem 3.5.1 we have $\lim_{x \to \infty} (f(x) + g(x)) = a + b$ and hence $\lim_{n \to \infty} (a_n + b_n) = a + b$. Thus (I) holds and similarly (II) and (III) hold. In order to ensure that (IV) holds we have to assume additionally that $g(x) \neq 0$. Hence, to complete the proof it suffices to prove the following theorem.

4.1.5 Theorem *For every sequence* $\{a_n\}$, *there exists a function* f *defined on* $[1, \infty)$ *such that* $f(n) = a_n$, *for all* $n \in \mathbb{N}$, *and such that* $\lim_{n \to \infty} a_n = a$ *if and only if* $\lim_{x \to \infty} f(x) = a$. *If, additionally,* $a_n \neq 0$ *for all* $n \in \mathbb{N}$, *then there exists a function* f *with the above properties such that* $f(x) \neq 0$ *for all* $x \in [1, \infty)$.

Proof. It suffices to put $f(x) = a_{[x]}$ where $[x]$ denotes, as usual, the greatest natural number less than or equal to x.

Using Theorem 4.1.5 we immediately obtain from 3.6.1 and 3.6.2 the following two theorems:

4.1.6 Theorem (on preserving inequality) *If* $a_n \leq b_n$ *for all* $n \in \mathbb{N}$, *then* $\lim_{n \to \infty} a_n \leq \lim_{n \to \infty} b_n$, *whenever the limits exist.*

4.1.7 Theorem (on three sequences) *If* $a_n \leq b_n \leq c_n$ *for all* $n \in \mathbb{N}$ *and* $\lim_{n \to \infty} a_n = \lim_{n \to \infty} c_n = k$, *then* $\lim_{n \to \infty} b_n = k$.

Theorem 4.1.5 reduces theorems on sequences to the corresponding theorems on functions. This is useful because we do not have to repeat similar arguments.

—Would it be possible to proceed in the opposite direction: first prove theorems on sequences and then transfer them to functions?

—Actually, this is a very common approach. One says that k is the limit of a function f at x_0 if for every sequence $x_1, x_2, \ldots$ convergent to x_0 the sequence $f(x_1), f(x_n), \ldots$ converges to k; that is, if $\lim_{n\to\infty} x_n = x_0$ then $\lim_{n\to\infty} f(x_n) = k$. This definition is attributed to Eduard Heine (1821–1881). The sequences used in the definition are sometimes called *Heine sequences*.

—Which approach is better, starting from functions or starting from sequences?

—It is only a matter of taste. The definitions are equivalent. The use of Heine's approach is sometimes very helpful, for example in the proof of 7.15.1. Sequences can also be used to define the limit at infinity. In that case we use sequences tending to infinity.

4.1.8 Definition   We say that a *sequence* $\{x_n\}$ *tends to infinity,* and write $x_n \to \infty$ or $\lim_{n\to\infty} x_n = \infty$, if for every number K there is an index n_0 such that $x_n > K$ for all $n > n_0$.

It is important to remember that a sequence tending to infinity does not have a limit in the sense of Definition 4.1.1.

Exercises 4.1

1. Use Definition 4.1.1 to prove the following:

    (a) $\lim_{n\to\infty} \dfrac{1}{n^2} = 0$,

    (b) $\lim_{n\to\infty} \dfrac{n+1}{n+2} = 1$,

    (c) $\left\{ \dfrac{n^2+1}{n+1} \right\}$ diverges,

    (d) $\{(-1)^n\}$ diverges,

    (e) $\lim_{n\to\infty} \dfrac{1}{n^2} \neq 10$,

    (f) $\lim_{n\to\infty} \dfrac{2n^2+1}{1+n+n^2} \neq 1$,

2. Prove that $\lim_{n\to\infty} \dfrac{1}{n!} = 0$ (where $n! = 1 \cdot 2 \cdots n$).

3. Prove that the sequence $a, a, a, \ldots$ converges to a.

4. Decide whether the following are possible:

    (a) $\{a_n\}$ converges, $\{b_n\}$ diverges, and $\{a_n + b_n\}$ converges,

    (b) $\{a_n\}$ and $\{b_n\}$ diverge, but $\{a_n + b_n\}$ converges,

    (c) $\{a_n\}$ converges, $\{b_n\}$ diverges, and $\{a_n b_n\}$ converges,

    (d) $\{a_n\}$ and $\{b_n\}$ diverge, but $\{a_n b_n\}$ converges.

5. Is it possible to define a continuous function f satisfying the conditions in Theorem 4.1.5?

6. True or false? If $a_n < b_n$ for all $n \in \mathbb{N}$, then $\lim_{n\to\infty} a_n < \lim_{n\to\infty} b_n$, whenever the limits exist.

7. Prove that if $\lim_{n\to\infty} a_n = a$, then $\lim_{n\to\infty} |a_n| = |a|$.

8. Prove that $\lim_{n\to\infty} a_n = 0$ if and only if $\lim_{n\to\infty} |a_n| = 0$.

9. Give an example of a bounded sequence that is not convergent.

10. True or false? If $0 \le a_n \le b_n$ for all $n \in \mathbb{N}$ and $\{b_n\}$ converges then $\{a_n\}$ converges.

11. Prove that if $\lim_{n \to \infty} a_n = a$ and $\lim_{n \to \infty} b_n = a$, then the sequence $a_1, b_1, a_2, b_2, a_3, b_3, \ldots$ converges to a.

12. Prove that $\lim_{x \to x_0} f(x) = k$ if and only if $\lim_{n \to \infty} f(x_n) = k$ for every sequence $x_1, x_2, \ldots$ in the domain of f such that $\lim_{n \to \infty} x_n = x_0$.

13. Prove that $\lim_{x \to \infty} f(x) = k$ if and only if $\lim_{n \to \infty} f(x_n) = k$ for every sequence $x_1, x_2, \ldots$ in the domain of f such that $\lim_{n \to \infty} x_n = \infty$.

4.2 MONOTONE SEQUENCES AND SUBSEQUENCES

A sequence $a_1, a_2, \ldots$ is called *nondecreasing* if $a_n \le a_{n+1}$ for every $n \in \mathbb{N}$. Similarly, $\{a_n\}$ is called *nonincreasing* if $a_n \ge a_{n+1}$ for every $n \in \mathbb{N}$. Both nondecreasing and nonincreasing sequences are called *monotone*. A constant sequence is nonincreasing and at the same time nondecreasing.

4.2.1 Theorem   *Every bounded monotone sequence is convergent.*

The proof follows from 3.23.3 and 4.1.5.

If we remove terms from a sequence but let infinitely many terms remain, the remaining terms form a new sequence, which is called a subsequence of the initial sequence. This intuitive definition can be formulated precisely as follows.

4.2.2 Definition   Let $\{a_n\}$ be a sequence and let p_n be an increasing sequence of natural numbers. The sequence $a_{p_1}, a_{p_2}, a_{p_3}, \ldots$ is called a *subsequence* of $\{a_n\}$.

The sequences $\{a_{2n}\}$, $\{a_{2n-1}\}$, and $\{a_{m+n}\}$ (where m is a fixed natural number), are subsequences of $\{a_n\}$. Similarly,

$$\frac{1}{2}, \frac{1}{4}, \frac{1}{8}, \frac{1}{16}, \ldots,$$

$$\frac{1}{2}, \frac{1}{20}, \frac{1}{200}, \frac{1}{2000}, \ldots$$

are subsequences of the sequence $\frac{1}{2}, \frac{1}{3}, \frac{1}{4}, \frac{1}{5}, \ldots$.

4.2.3 Theorem   *A subsequence of a convergent sequence converges to the same limit.*

Proof.   Let $\lim_{n \to \infty} a_n = a$ and let $\{a_{p_n}\}$ be a subsequence of $\{a_n\}$. For every $\varepsilon > 0$ there is a number n_0 such that $|a_n - a| < \varepsilon$ for all $n > n_0$. Since $p_n \ge n$, which is easily checked by induction, we have $|a_{p_n} - a| < \varepsilon$ for all $n > n_0$, proving the theorem.

The next theorem is usually credited to two mathematicians: Bernard Bolzano (1781–1848) and Karl Weierstrass. It describes one of the fundamental properties of the real numbers.

4.2.4 Theorem (Bolzano–Weierstrass theorem) *Every bounded sequence contains a convergent subsequence.*

Proof.   Let $\{x_n\}$ be a bounded sequence. If $\{x_n\}$ contains a constant subsequence then the theorem is obviously satisfied. Assume then that $\{x_n\}$ does not contain a constant subsequence. Let, for every $n \in \mathbb{N}$, y_n be the least upper bound of the set $\{x_n, x_{n+1}, \ldots\}$. The sequence $\{y_n\}$ is nonincreasing and bounded. Hence, by Theorem 4.2.1, it converges to a limit y. Denote by z_n the first term of the sequence $x_n, x_{n+1}, \ldots$ such that

$$|z_n - y_n| < \frac{1}{n}.$$

Since every term of the sequence $\{x_n\}$ can appear only a finite number of times in the sequence $\{z_n\}$, there is a subsequence $\{z_{p_n}\}$ of $\{z_n\}$ that is simultaneously a subsequence of $\{x_n\}$. We have

$$|z_{p_n} - y| \le |z_{p_n} - y_{p_n}| + |y_{p_n} - y| < \frac{1}{p_n} + |y_{p_n} - y| \to 0.$$

Therefore $\{z_{p_n}\}$ is a convergent subsequence of $\{x_n\}$.

Exercises 4.2

1. Check whether the following sequences are monotone.

    (a) $\left\{ \dfrac{1}{n^2} \right\}$
    (b) $\left\{ \dfrac{n}{n+1} \right\}$
    (c) $\left\{ \dfrac{n^4}{n!} \right\}$

2. Define

    $$a_1 = 1; \qquad a_{n+1} = \frac{3a_n + 4}{2a_n + 3} \quad \text{for } n \ge 2.$$

    Show that $\{a_n\}$ converges.

3. Define

    $$a_1 = 1; \qquad a_{n+1} = \frac{a_n + 2}{a_n + 1} \quad \text{for } n \ge 2.$$

    Show that $\{a_n\}$ converges.

4. Prove that $\{(1 + 1/n)^n\}$ is a convergent sequence.

5. Prove that every Cauchy sequence converges (see Section 2.13 for the definition of a Cauchy sequence).

4.3 GEOMETRIC SEQUENCES

The sequence $\{q^n\}$, where q is an arbitrary real number, is called a *geometric sequence*. An example of such a sequence is

$$\frac{1}{3}, \frac{1}{9}, \frac{1}{27}, \frac{1}{81}, \ldots,$$

which is $\{1/3^n\}$. The choice $q = 1$ produces the constant sequence $1, 1, 1, \ldots$, which converges to 1. If $q = -1$ the geometric sequence is $-1, 1, -1, 1, \ldots$, which is not convergent. It contains a subsequence convergent to 1 and a subsequence convergent to -1.

4.3.1 Theorem   *If* $|q| < 1$, *then* $\lim_{n \to \infty} q^n = 0$.

Proof.   If $0 < q < 1$, then the sequence $\{q^n\}$ is nonincreasing and bounded: $q^{n+1} = qq^n < q^n$ and $0 < q^n < 1$. Hence, by Theorem 4.2.1, it converges to some limit k. From the equation $q^{n+1} = qq^n$, we obtain

$$k = \lim_{n \to \infty} q^{n+1} = \lim_{n \to \infty} qq^n = q \lim_{n \to \infty} q^n = qk,$$

because $\{q^{n+1}\}$ is a subsequence of $\{q^n\}$. The only k that satisfies the last equality is $k = 0$. This proves the theorem for $0 < q < 1$.

If $-1 < q < 0$, then $0 < |q| < 1$, and thus $\lim_{n \to \infty} |q|^n = 0$, by the first part of the proof. This implies that $\lim_{n \to \infty} q^n = 0$, by Theorem 4.1.3, because $|q^n| = |q|^n$.

Since obviously $\lim_{n \to \infty} q^n = 0$ if $q = 0$, the proof is complete.

Note that if $q > 1$ then $\lim_{n \to \infty} q^n = \infty$. If $q < -1$, then the sequence $\{q^n\}$ oscillates between larger and larger positive and negative values.

Exercises 4.3

1. Show that $\lim_{n \to \infty} q^{n^2} = 0$ if $|q| < 1$.
2. Find the limit $\lim_{n \to \infty} (2^n - n^2)/(n^2 \cdot 2^n)$.
3. Find the limit $\lim_{n \to \infty} 2 \cdot 3^{1+n} \cdot 5^{1-n}$.
4. Find the limit $\lim_{n \to \infty} (-2)^{3n} \cdot 3^{-2n}$.

4.4 GEOMETRIC SERIES

Let us consider the sequence $\{s_n\}$ defined by induction as follows:

(1) $s_1 = 1;$ $s_{n+1} = s_n + q^n,$ for $n \in \mathbb{N}$ and some $q \in \mathbb{R}$.

Thus we have

$$s_2 = 1 + q,$$
$$s_3 = 1 + q + q^2,$$
$$s_4 = 1 + q + q^2 + q^3,$$

and generally

$$s_{n+1} = 1 + q + q^2 + \cdots + q^n.$$

The sequence $\{s_n\}$ is called a *geometric series*. Although the last expression has a clear intuitive meaning, the rigorous definition is (1).

4.4.1 Theorem   *If $|q| < 1$, then the sequence defined by (1) is convergent and*

$$\lim_{n \to \infty} s_n = \frac{1}{1 - q},$$

which can be written as

$$1 + q + q^2 + q^3 + \cdots = \frac{1}{1 - q}.$$

Proof.   It follows by induction that

$$(1 - q)s_n = 1 - q^n.$$

Dividing both sides by $1 - q$ we obtain

$$s_n = \frac{1 - q^n}{1 - q},$$

and hence

$$\lim_{n \to \infty} s_n = \lim_{n \to \infty} \frac{1 - q^n}{1 - q} = \frac{1}{1 - q},$$

by 4.3.1.

The above theorem has a simple geometric interpretation; see Fig. 4.1. It is easily seen that the triangles AOB and CBD are similar. Hence $AO/BA = BC/CD$; that is,

$$\frac{1 + q + q^2 + \cdots}{1} = \frac{1}{1 - q}.$$

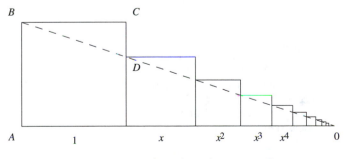

Figure 4.1 $1 + q + q^2 + q^3 + \cdots = \frac{1}{1-q}$.

Exercises 4.4 ──

 1. Let $k \in \mathbb{N}$. Show that $1 + q^k + q^{2k} + q^{3k} + \cdots = 1/(1 - q^k)$ for every $|q| < 1$.
 2. Show that $1 - q + q^2 - q^3 + \cdots = 1/(1 + q)$ for every $|q| < 1$.

──

4.5 GENERAL CONCEPT OF A SERIES

The geometric series is obtained from the sequence $1, q, q^2, q^3, \ldots$ by building sums s_n defined by Equation (1) of the preceding section. Similarly, from any sequence $\{a_n\}$ we can build a series by using the formulas

(1) $s_1 = a_1;\qquad s_{n+1} = s_n + a_{n+1}\quad$ for all $n \in \mathbb{N}$.

The sums s_n are called the *partial sums* of the sequence $\{a_n\}$.

For example, if $a_n = 1/n(n + 1)$, then it can be easily shown by induction that $s_n = 1 - 1/(n + 1)$. In this case the series can be written as

$$\frac{1}{1 \cdot 2}, \quad \frac{1}{1 \cdot 2} + \frac{1}{2 \cdot 3}, \quad \frac{1}{1 \cdot 2} + \frac{1}{2 \cdot 3} + \frac{1}{3 \cdot 4}, \cdots$$

or equivalently

$$\frac{1}{2}, \quad \frac{2}{3}, \quad \frac{3}{4}, \ldots.$$

—What is the definition of a series?

—It is rather difficult to give a logical definition of a series. If we were to speak of a series we would simultaneously think of both the sequence

$$a_1, \ a_2, \ a_3, \ldots$$

and the sequence of its partial sums

$$s_1, \ s_2, \ s_3, \ldots.$$

—So the theory of series is a part of the theory of sequences—sequences of particular form. Could we say that a series is a sequence of partial sums?

—This would not solve the problem. Any sequence $a_1, a_2, a_3, \ldots$ can be considered a sequence of partial sums: it suffices to take

$$b_1 = a_1;\qquad b_n = a_n - a_{n-1}\quad \text{for } n \geq 2.$$

Then $b_1 + \cdots + b_n = a_n$.

—And what is the definition in other books?

—Their authors usually speak of series without giving a precise definition. The prominent specialist in the theory of series Konrad Knopp (1882–1957) said that a series is a sequence that is defined in a special way: one starts with a sequence $a_1, \ a_2, \ a_3, \ldots$, and then constructs a new sequence by letting

$$s_1 = a_1, \qquad s_2 = a_1 + a_2, \qquad s_3 = a_1 + a_2 + a_3, \ldots.$$

According to Knopp a series is a symbol $a_1 + a_2 + a_3 + \cdots$ or $\sum_{k=1}^{\infty} a_k$.

—We never define anything to be a symbol. Is it correct?

—It may be correct, but such definitions are not used. Knopp wrote his famous book *Theory and Applications of Infinite Series* in 1921. Since that time mathematics has become more rigorous and it is not common to define mathematical objects by symbols.

—Then what actually is a series?

—If one insists on having a definition, then we can say that a series is a pair of sequences $a_1, a_2, a_3, \ldots$ and $s_1, s_2, s_3, \ldots$ connected by equalities (1).

—What is the use of such a definition?

—None, but we have a definition that is adequate and logically correct. In order to understand convergence of series all we really need is the concept of the sum of a sequence. The sum of a sequence $a_1, a_2, a_3, \ldots$ is defined as $\lim_{n \to \infty} s_n$, where the s_n are defined by (1). As in the case of limits of sequences, some sequences have sums, other sequences do not have sums. The expression "series converges" simply means that the sequence of partial sums has a limit. Similarly, the expression "series diverges" means that the sequence of partial sums does not have a limit.

The *sum of an infinite sequence* $\{a_n\}$ is denoted by $\sum_{k=1}^{\infty} a_k$. The partial sums are often written in the form

$$ s_n = \sum_{k=1}^{n} a_k. $$

Thus we have

$$ \sum_{k=1}^{\infty} a_k = \lim_{n \to \infty} s_n = \lim_{n \to \infty} \sum_{k=1}^{n} a_k. $$

Since $\sum_{k=1}^{\infty} a_k$ denotes the sum, it is a number. The symbol $\sum_{k=1}^{\infty} (-1)^k$ is meaningless, because the sequence $-1, 1, -1, 1, \ldots$ has no sum. Although the sequence $1, 1, 1, \ldots$ has no sum, we sometimes write $\sum_{k=1}^{\infty} 1 = \infty$. This agrees with Definition 4.1.8, because the nth partial sum of the series equals n, and $\lim_{n \to \infty} n = \infty$. However, this series is divergent, as is the sequence $1, 2, 3, \ldots$.

To denote the series formed from the sequence $a_1, a_2, a_3, \ldots$ we will use the symbol $\left\{ \sum_{k=1}^{n} a_k \right\}$. The braces are used to distinguish the series from its nth partial sum, which is $\sum_{k=1}^{n} a_k$. This is in accordance with our notation for sequences: $\{a_n\}$ denotes the sequence, while a_n denotes its nth term.

We often use the following intuitive notation

$$ a_1 + a_2 + a_3 + \cdots = s. $$

This means that the series $\left\{ \sum_{k=1}^{n} a_k \right\}$ converges and $\sum_{k=1}^{\infty} a_k = s$.

4.5.1 Theorem   *If the series $\left\{ \sum_{k=1}^{n} a_k \right\}$ converges, then $\lim_{n \to \infty} a_n = 0$.*

Proof.   Let $s_n = \sum_{k=1}^{n} a_k$, $n \in \mathbb{N}$. Convergence of the series $\left\{ \sum_{k=1}^{n} a_k \right\}$ means that $\lim_{n \to \infty} s_n = s$ for some $s \in \mathbb{R}$. Therefore, for every $\varepsilon > 0$ there exists a number n_0 such that for $n > n_0$ we have

$$ |s_n - s| < \frac{1}{2} \varepsilon. $$

Thus for $n > n_0 + 1$ we have

$$|a_n| = |s_n - s_{n-1}| \le |s_n - s| + |s - s_{n-1}| < \frac{1}{2}\varepsilon + \frac{1}{2}\varepsilon = \varepsilon.$$

Since ε is an arbitrary small number, the theorem is proved.

The converse of Theorem 4.5.2 is not true. For example, the sequence $1, \frac{1}{2}, \frac{1}{3}, \frac{1}{4}, \ldots$ converges to 0, but the series $\{\sum_{k=1}^{n} 1/k\}$ diverges. Divergence of this series follows from the so-called *integral test* (Theorem 7.20.1). It can also be proved directly by considering partial sums of the form

$$1 + \frac{1}{2},$$

$$1 + \frac{1}{2} + \left(\frac{1}{3} + \frac{1}{4}\right),$$

$$1 + \frac{1}{2} + \left(\frac{1}{3} + \frac{1}{4}\right) + \left(\frac{1}{5} + \frac{1}{6} + \frac{1}{7} + \frac{1}{8}\right),$$

$$1 + \frac{1}{2} + \left(\frac{1}{3} + \frac{1}{4}\right) + \left(\frac{1}{5} + \frac{1}{6} + \frac{1}{7} + \frac{1}{8}\right) + \left(\frac{1}{9} + \frac{1}{10} + \frac{1}{11} + \frac{1}{12} + \frac{1}{13} + \frac{1}{14} + \frac{1}{15} + \frac{1}{16}\right),$$

and so on.

4.5.2 Theorem   *If the series $\{\sum_{k=1}^{n} a_k\}$ converges, then for any constant $c \in \mathbb{R}$ the series $\{\sum_{k=1}^{n} ca_k\}$ converges and*

$$(2) \qquad \sum_{k=1}^{\infty} ca_k = c \sum_{k=1}^{\infty} a_k.$$

Proof.   By an easy induction we obtain

$$(3) \qquad \sum_{k=1}^{n} ca_k = c \sum_{k=1}^{n} a_k.$$

Hence, by letting $n \to \infty$, we get (2). $\blacksquare$

The phrase "by letting $n \to \infty$" does not mean that we merely replace n by ∞. It is a short way of indicating that we take the limit as $n \to \infty$ on both sides of (3). This is valid only if the limits exist. The detailed proof of (2) would require the following steps:

$$\sum_{k=1}^{\infty} ca_k = \lim_{n \to \infty} \sum_{k=1}^{n} ca_k = \lim_{n \to \infty}\left(c \sum_{k=1}^{n} a_k\right) = c \lim_{n \to \infty} \sum_{k=1}^{n} a_k = c \sum_{k=1}^{\infty} a_k.$$

The first and last equalities follow from the definition of the sum of a series, the second equality follows from (3), and the third equality follows from 4.1.4. Whenever the phrase "by letting $n \to \infty$" is used, we should check that the argument is justified.

4.5.3 Theorem   *If the series $\{\sum_{k=1}^{n} a_k\}$ and $\{\sum_{k=1}^{n} b_k\}$ are convergent then the series $\{\sum_{k=1}^{n}(a_k + b_k)\}$ is convergent and*

(4)
$$\sum_{k=1}^{\infty}(a_k + b_k) = \sum_{k=1}^{\infty} a_k + \sum_{k=1}^{\infty} b_k.$$

Proof.   Since $\sum_{k=1}^{n}(a_k + b_k) = \sum_{k=1}^{n} a_k + \sum_{k=1}^{n} b_k$ for every $n \in \mathbb{N}$, we obtain (4) by letting $n \to \infty$.

Exercises 4.5

1. Show that

$$\sum_{k=1}^{n} \frac{1}{k(k + 1)} = 1 - \frac{1}{k + 1}.$$

2. Find a sequence for which the sequence of partial sums is $1, \frac{1}{2}, \frac{1}{3}, \frac{1}{4}, \frac{1}{5}, \ldots$.

3. Test the following for convergence. Find the sum if convergent.

   (a) $\left\{\sum_{k=1}^{n} \frac{k^2}{k^2 + 1}\right\}$ (b) $\left\{\sum_{k=1}^{n}\left(1 + \frac{1}{k}\right)^k\right\}$

   (c) $\left\{\sum_{k=1}^{n} \frac{1}{k^2 + 2k}\right\}$ (d) $\left\{\sum_{k=1}^{n} \frac{1}{k^2 + 4k + 3}\right\}$

   (e) $\left\{\sum_{k=1}^{n}(-2)^{3k} \cdot 3^{-2k}\right\}$ (f) $\left\{\sum_{k=1}^{n} 2 \cdot 3^{1+k} \cdot 5^{1-k}\right\}$

4. Show that the series $\{\sum_{k=1}^{n} 1/k\}$ diverges.

5. Show whether the following is possible: $\{\sum_{k=1}^{n} a_k\}$ converges, $\{\sum_{k=1}^{n} b_k\}$ diverges, and $\{\sum_{k=1}^{n}(a_k + b_k)\}$ converges.

6. Show whether the following is possible: Both $\{\sum_{k=1}^{n} a_k\}$ and $\{\sum_{k=1}^{n} b_k\}$ diverge, but $\{\sum_{k=1}^{n}(a_k + b_k)\}$ converges.

4.6 CONVERGENT AND ABSOLUTELY CONVERGENT SERIES

A proof that a series is convergent can often be carried out by comparing it with some other series which is known to be convergent.

4.6.1 Theorem (comparison test)   *If $|a_n| \leq b_n$ for all $n \in \mathbb{N}$ and the series $\{\sum_{k=1}^{n} b_k\}$ is convergent, then the series $\{\sum_{k=1}^{n} a_k\}$ is convergent and $\left|\sum_{n=1}^{\infty} a_n\right| \leq \sum_{n=1}^{\infty} b_n$.*

Proof.   Let $\{p_n\}$ be the sequence obtained from $\{a_n\}$ by replacing all negative terms by 0: $p_n = \frac{1}{2}(|a_n| + a_n)$. Similarly, let $\{q_n\}$ be the sequence that is obtained by replacing all positive terms by 0 and changing the sign of the remaining terms: $q_n = p_n - a_n$. Then

(1) $$a_n = p_n - q_n.$$

Moreover, since $|a_n| \le b_n$, we have

(2) $$0 \le p_n \le b_n \quad \text{and} \quad 0 \le q_n \le b_n.$$

Denote, for $n = 1, 2, 3, \ldots,$

$$s_n = \sum_{k=1}^{n} a_k, \qquad t_n = \sum_{k=1}^{n} b_k, \qquad u_n = \sum_{k=1}^{n} p_k, \quad \text{and} \quad v_n = \sum_{k=1}^{n} q_k.$$

From (1) and (2), by induction, we get

(3) $$s_n = u_n - v_n, \qquad 0 \le u_n \le t_n, \quad \text{and} \quad 0 \le v_n \le t_n.$$

Since, by the assumption of the theorem, the sequence $\{t_n\}$ is convergent, it is bounded; that is, there is a number M such that $t_n \le M$ for all $n \in \mathbb{N}$. Hence

$$0 \le u_n \le M \quad \text{and} \quad 0 \le v_n \le M.$$

Moreover, the sequences $\{u_n\}$ and $\{v_n\}$ are nondecreasing (as sequences of partial sums of sequences with nonnegative terms). Therefore, by Theorem 4.2.1, they are convergent, implying, in view of (1) and Theorem 4.5.3, that the series $\{\sum_{k=1}^{n} a_k\}$ is convergent.

Moreover, from (3) it follows that $-t_n \le s_n \le t_n$ and, as $n \to \infty$, we derive

$$-\sum_{k=1}^{\infty} b_k \le \sum_{k=1}^{\infty} a_k \le \sum_{k=1}^{\infty} b_k,$$

which completes the proof.

4.6.2 Definition   A series $\{\sum_{k=1}^{n} a_k\}$ is *absolutely convergent* if the series $\{\sum_{k=1}^{n} |a_k|\}$ is convergent.

Obviously, every convergent series of non-negative terms is absolutely convergent. Note that the series $\{\sum_{k=1}^{n} a_k\}$ in Theorem 4.6.1 is actually absolutely convergent.

Not every convergent series is absolutely convergent (see Exercise 4.6.4). On the other hand we have the following theorem.

4.6.3 Theorem   *Every absolutely convergent series converges.*

The proof follows from Theorem 4.6.1 by letting $b_n = |a_n|$.

—Suppose we take a convergent series $\{\sum_{k=1}^{n} a_k\}$. Then we put all the terms of the sequence $\{a_n\}$ into a sack, mix them thoroughly, and pick them at random one after another to form a new series. Will that new series converge? Is the sum going to be the same?

—Various things can happen. For instance, the series

$$\left\{ \sum_{k=1}^{n} \frac{(-1)^{k+1}}{k} \right\}$$

is convergent (see Theorem 4.36.1), but the sequence $1, -\frac{1}{2}, \frac{1}{3}, -\frac{1}{4}, \ldots$ can be rearranged to form a divergent series. We can pick from it first only positive numbers so that the sum reaches an arbitrary large number M. Taking next only negative terms, we can come up with negative partial sums and reach an arbitrary large negative number. Continuing in this manner we can form a sequence whose partial sums "oscillate between ∞ and $-\infty$," so that it is not convergent. It is still more unexpected that by properly rearranging the terms of the above sequence we can make the series converge to any given number K. This is possible because the series

$$\left\{ \sum_{k=1}^{\infty} \frac{(-1)^{k+1}}{k} \right\}$$

is not absolutely convergent. A convergent series that is not absolutely convergent is called *conditionally convergent*.

—And what happens if the series is absolutely convergent?

—Then no rearrangement affects the absolute convergence or the sum of the series. This property can be generalized as follows.

If we are given any finite number n of series absolutely convergent to sums $k_1, \ldots, k_n$ and we put all terms of the sequences into a common sack, mix them up, and then form a new series by drawing out consecutive terms at random such that all the terms are used, then the series obtained is absolutely convergent to the sum $k_1 + \cdots + k_n$. This result can be still strengthened by admitting a whole infinite sequence of absolutely convergent series. However, mixing is not a mathematical concept and we must make it more precise. What does a "sack" of terms mean? This should mean that we do not care about the order of terms. But in a sack there may exist identical terms. That is why we cannot identify the contents of a sack with a set. The difficulty disappears if we carefully distinguish between the set of all numbers that appear in a sequence and the terms of the sequence. A term of a sequence is a number with an index showing the place of that number in the sequence. If the same number, say 1, appears twice in a sequence, then it is counted as two different terms of the sequence. The exact definition of mixing can be based on the concept of rearrangement.

4.6.4 Definition By a *rearrangement* of the sequence of natural numbers $1, 2, 3, \ldots$ we mean an infinite matrix

(4)
$$\begin{array}{llll} p_{1,1}, & p_{1,2}, & p_{1,3}, \cdots \\ p_{2,1}, & p_{2,2}, & p_{2,3}, \cdots \\ \cdots \end{array}$$

that includes all natural numbers and includes each number only once. In other words:

For every $j \in \mathbb{N}$ there exist $i, k \in \mathbb{N}$ such that $j = p_{i,k}$ and $p_{i,k} \neq p_{m,n}$ whenever $i \neq m$ or $k \neq n$.

The number of rows in the matrix can be finite or infinite. In particular it may have only one row, which means that we have only one sequence in (4).

4.6.5 Definition   We say that a series $\{\sum_{k=1}^{n} a_k\}$ is *composed* of series

(5)
$$\left\{\sum_{k=1}^{n} b_{1,k}\right\}, \left\{\sum_{k=1}^{n} b_{2,k}\right\}, \left\{\sum_{k=1}^{n} b_{3,k}\right\}, \dots$$

if there is a rearrangement of the sequence $1, 2, 3, \dots$ (with the same number of rows as the number of series in (5)) such that $b_{i,j} = a_{p_{i,j}}$.

4.6.6 Theorem   *Let a series $\{\sum_{k=1}^{n} a_k\}$ be composed of an infinite number of absolutely convergent series*

(6)
$$\left\{\sum_{k=1}^{n} b_{1,k}\right\}, \left\{\sum_{k=1}^{n} b_{2,k}\right\}, \left\{\sum_{k=1}^{n} b_{3,k}\right\}, \dots,$$

and let

$$\sum_{k=1}^{\infty} b_{m,k} = B_m \quad \text{and} \quad \sum_{k=1}^{\infty} |b_{m,k}| = C_m.$$

If the series $\{\sum_{k=1}^{n} C_k\}$ is convergent, then the series $\{\sum_{k=1}^{n} a_k\}$ is absolutely convergent and

(7)
$$\sum_{k=1}^{\infty} a_k = \sum_{k=1}^{\infty} B_k.$$

Proof.   Let (4) be the rearrangement of the sequence $1, 2, 3, \dots$, for which $b_{i,j} = a_{p_{i,j}}$. Clearly, for any natural number n there exists another natural number m such that among those $p_{i,j}$ for which $i, j \le m$ there appear all the numbers $1, \dots, n$. Then we have

$$|a_1| + \dots + |a_n| \le (|b_{1,1}| + \dots + |b_{1,m}|) + \dots + (|b_{m,1}| + \dots + |b_{m,m}|).$$

Since the sum on the right is always less than or equal to $\sum_{k=1}^{\infty} C_k$, the series $\{\sum_{k=1}^{n} a_k\}$ converges absolutely.

The absolute convergence of the series $\{\sum_{k=1}^{n} B_k\}$ follows immediately from the inequalities $|B_n| \le C_n$. Therefore both series $\{\sum_{k=1}^{n} a_k\}$ and $\{\sum_{k=1}^{n} B_k\}$ converge, by Theorem 4.6.3.

To prove (7) we shall show that

(8)
$$(B_1 + \dots + B_n) - (a_1 + \dots + a_n)$$

converges to 0 as $n \to \infty$. Let ε be a positive number. Since $\{\sum_{k=1}^{n} a_k\}$ converges absolutely there is a natural number m such that

$$|a_{m+1}| + |a_{m+2}| + |a_{m+3}| + \dots < \varepsilon.$$

We choose r such that $p_{i,j} > m$ for $i, j > r$ and such that among the $p_{i,j}$ for which $i, j \le r$ there appear all the numbers $1, \dots, m$. Let

$$s_{i,n} = b_{i,1} + \dots + b_{i,n}.$$

Then for $n > r$ the difference

$$(s_{1,n} + \dots + s_{n,n}) - (a_1 + \dots + a_n)$$

contains, after canceling equal terms, only those terms $b_{i,j}$ for which $i, j > m$. Hence

(9)
$$\left|(s_{1,n} + \cdots + s_{n,n}) - (a_1 + \cdots + a_n)\right| < \varepsilon \quad \text{for} \quad n > r.$$

Similarly, for every i and every $n > r$, the right-hand side of the equality

$$B_i - s_{i,n} = b_{i,n+1} + b_{i,n+2} + b_{i,n+3} + \cdots$$

contains only terms $a_{p_{i,j}}$ with $p_{i,j} > m$, and therefore

(10)
$$\left|B_1 - s_{1,n}\right| + \cdots + \left|B_n - s_{n,n}\right| < \varepsilon \quad \text{for} \quad n > r.$$

From (9) and (10) we get

$$\left|(B_1 + \cdots + B_n) - (a_1 + \cdots + a_n)\right| < 2\varepsilon \quad \text{for } n > r.$$

Since ε is arbitrary, the proof is complete.

The above theorem is formulated for an infinite number of series in (6), but it is rather obvious that it is also true for a finite number. This can be formally obtained from the infinite case by letting $a_{i,j} = 0$ for all $i > m$ (and all $j \in \mathbb{N}$), where m is the number of series.

Equality (7) can be written in a more intuitive but less precise way as

$$a_1 + a_2 + a_3 + \cdots = b_{1,1} + b_{1,2} + b_{1,3} + \cdots + b_{2,1} + b_{2,2} + b_{2,3} + \cdots$$
$$+ b_{n,1} + b_{n,2} + b_{n,3} + \cdots.$$

Since we are adding the same numbers on the left and on the right, although in a different order, the equality may seem obvious. We know that the algebraic operation addition has the property that the sum does not depend on the order in which the terms are summed. However, the operation of adding infinitely many numbers is not an algebraic operation, so the rules of addition do not automatically apply to series. The theory of convergent series is largely devoted to study of which properties of algebraic operations on numbers extend to series, and under what conditions. The above theorem, for instance, shows that absolute convergence is crucial for the "reordering" (or "regrouping") of infinite sums. A good understanding of this and the following theorem will be extremely important in Chapter 7.

The following theorem is somewhat converse to 4.6.6.

4.6.7 Theorem *Let an absolutely convergent series $\{\sum_{k=1}^{n} a_k\}$ be composed of the series*

(11)
$$\left\{\sum_{k=1}^{n} b_{1,k}\right\}, \left\{\sum_{k=1}^{n} b_{2,k}\right\}, \left\{\sum_{k=1}^{n} b_{3,k}\right\}, \ldots$$

Then each of the series in (11) is absolutely convergent and

(12)
$$\sum_{k=1}^{\infty} a_k = \sum_{i=1}^{\infty} \left(\sum_{j=1}^{\infty} b_{i,j}\right)$$

(that is, $a_1 + a_2 + \cdots = (b_{1,1} + b_{1,2} + \cdots) + (b_{2,1} + b_{2,2} + \cdots) + \cdots$).

Proof.   To prove the absolute convergence of series (11) it suffices to note that for any indices i and n we have

$$|b_{i,1}| + \cdots + |b_{i,n}| \leq \sum_{k=1}^{\infty} |a_k|.$$

The equality (12) follows from 4.6.6.

Exercises 4.6

1. Prove convergence of the following series:

    (a) $\left\{ \displaystyle\sum_{k=1}^{n} \frac{1}{2^k + \sqrt{k}} \right\}$;

    (b) $\left\{ \displaystyle\sum_{k=1}^{n} \frac{(-1)^k}{k - 3^k} \right\}$.

2. Prove that if $\{\sum_{k=1}^{n} a_k\}$ and $\{\sum_{k=1}^{n} b_k\}$ converge absolutely, then $\{\sum_{k=1}^{n} a_k b_k\}$ converges absolutely. Does this imply that $\sum_{k=1}^{\infty} a_k b_k = \left(\sum_{k=1}^{\infty} a_k\right) \cdot \left(\sum_{k=1}^{\infty} b_k\right)$?

3. Prove that if the series $\{\sum_{k=1}^{n} a_k\}$ converges absolutely and the sequence $\{b_n\}$ is bounded, then the series $\{\sum_{k=1}^{n} a_k b_k\}$ converges absolutely.

4. Prove that the series formed from the sequence $1, -1, \frac{1}{2}, -\frac{1}{2}, \frac{1}{3}, -\frac{1}{3}, \ldots$ is conditionally convergent.

5. Find a rearrangement $\{a_n\}$ of the sequence $1, -1, \frac{1}{2}, -\frac{1}{2}, \frac{1}{3}, -\frac{1}{3}, \ldots$ for which the series $\{\sum_{k=1}^{\infty} a_n\}$ diverges.

6. True or false? If the series $\{\sum_{k=1}^{n} a_k\}$ converges and the sequence $\{b_n\}$ is bounded, then the series $\{\sum_{k=1}^{n} a_k b_k\}$ converges.

7. Let $\{\sum_{k=1}^{n} a_k\}$ be a conditionally convergent series. Define

    $$p_n = \frac{1}{2}(|a_n| + a_n) \quad \text{and} \quad q_n = \frac{1}{2}(|a_n| - a_n).$$

    Prove that both series $\{\sum_{k=1}^{n} p_k\}$ and $\{\sum_{k=1}^{n} q_k\}$ diverge.

8. Let $\{\sum_{k=1}^{n} a_k\}$ be a conditionally convergent series and let $K \in \mathbb{R}$. Prove that there exists a rearrangement $\{b_k\}$ of $\{a_k\}$ such that $\sum_{n=1}^{\infty} b_n = K$.

9. Let $\{\sum_{k=1}^{n} a_k\}$ be an absolutely convergent series and let $\{b_n\}$ be a subsequence of $\{a_n\}$. Prove that the series $\{\sum_{k=1}^{n} b_k\}$ converges absolutely.

10. Let $\{\sum_{k=1}^{n} a_k\}$ be an absolutely convergent series and let, for all $n \in \mathbb{N}$, $b_n = a_{2n-1} + a_{2n}$. Prove that the series $\{\sum_{k=1}^{n} b_k\}$ is absolutely convergent and that $\sum_{k=1}^{\infty} a_k = \sum_{k=1}^{\infty} b_k$.

11. Let $\{\sum_{k=1}^{n} a_k\}$ be an absolutely convergent series and let, for all $n \in \mathbb{N}$, $b_n = a_{3n-2} + a_{3n-1} + a_{3n}$. Prove that the series $\{\sum_{k=1}^{n} b_k\}$ is absolutely convergent and that $\sum_{k=1}^{\infty} a_k = \sum_{k=1}^{\infty} b_k$.

12. Find $\sum_{n=1}^{\infty} n/2^n$.

13. Prove that if the limit $\lim_{n\to\infty} a_n/b_n$ exists (is finite) and is not 0, then $\{\sum_{k=1}^{n} a_k\}$ converges absolutely if and only if $\{\sum_{k=1}^{n} b_k\}$ converges absolutely.

14. True or false? If the limit $\lim_{n\to\infty} a_n/b_n$ exists (is finite) and is not 0, then $\{\sum_{k=1}^{n} a_k\}$ converges if and only if $\{\sum_{k=1}^{n} b_k\}$ converges.

15. Prove that $\{\sum_{k=1}^{n} a_k\}$ converges if and only if for every $\varepsilon > 0$ there exists $m \in \mathbb{N}$ such that

$$|a_m + a_{m+1} + \cdots + a_{m+p}| < \varepsilon \quad \text{for all } p \in \mathbb{N}.$$

16. Prove that if $\{\sum_{k=1}^{n} a_k\}$ converges absolutely, then for every $\varepsilon > 0$ there is a natural number m such that

$$|a_{m+1}| + |a_{m+2}| + |a_{m+3}| + \cdots < \varepsilon.$$

4.7 SUMMATION BETWEEN ARBITRARY LIMITS

In Section 4.5 we discussed the meanings of the symbols $\sum_{k=1}^{n}$ and $\sum_{k=1}^{\infty}$. More generally one can introduce the summation sign between arbitrary limits

$$\sum_{k=n+1}^{n+m} a_k;$$

it denotes the sum of the numbers a_k whose indices run from $k = n+1$ to $k = n+m$. The limiting indices $n + 1$ and $n + m$ may be arbitrary numbers with the only restriction $n + 1 \leq n + m$. The same sum can be written in a more intuitive way

$$\sum_{k=n+1}^{n+m} a_k = a_{n+1} + \ldots + a_{n+m}.$$

A precise definition of the above expressions can be based on the induction principle. One can also use the previously introduced symbol $\sum_{k=1}^{n} a_k$ and write

$$\sum_{k=n+1}^{n+m} a_k = \sum_{k=1}^{m} a_{n+k},$$

where n is arbitrary integer. For example, for $n = -1$ and $m = 1, 2, 3$ we obtain

$$\sum_{k=0}^{0} a_k = a_0, \qquad \sum_{k=0}^{1} a_k = a_0 + a_1, \qquad \sum_{k=0}^{2} a_k = a_0 + a_1 + a_2.$$

Similarly, we adopt the following definition

$$\sum_{k=n+1}^{\infty} a_k = \sum_{k=1}^{\infty} a_{n+k},$$

where n is an arbitrary integer. In particular, for $n = -1$, we have

$$\sum_{k=0}^{\infty} a_k = \sum_{k=1}^{\infty} a_{k-1} = a_0 + a_1 + a_2 + \cdots$$

The following property of infinite series is often useful:

4.7.1 *If either of the following two series*

(1)
$$\left\{ \sum_{k=m+1}^{m+n} a_k \right\} \quad \text{or} \quad \left\{ \sum_{k=m+q+1}^{m+q+n} a_k \right\}, \quad \text{where } q \in \mathbb{N},$$

is convergent, then the other series is also convergent and

(2)
$$\sum_{k=m+1}^{\infty} a_k = \sum_{k=m+1}^{m+q} a_k + \sum_{k=m+q+1}^{\infty} a_k.$$

Proof. Let

$$s_n = \sum_{k=m+1}^{m+n} a_k = \sum_{k=1}^{n} a_{m+k} \quad \text{and} \quad t_n = \sum_{k=m+q+1}^{m+q+n} a_k = \sum_{k=1}^{n} a_{m+q+k}.$$

In particular, $t_1 = a_{m+q+1}$. Hence $s_{q+1} = s_q + t_1$. By induction we get

$$s_{q+p} = s_q + t_p$$

for any $p \in \mathbb{N}$. By letting $p \to \infty$, the last equality becomes (2).

Note that 4.7.1 justifies the following (intuitively obvious) property

$$a_{n+1} + a_{n+2} + a_{n+3} + \cdots = (a_{n+1} + \cdots + a_{n+q}) + (a_{n+q+1} + a_{n+q+2} + \cdots).$$

4.8 DECIMAL FRACTIONS

Let $\{a_n\}$ be an infinite sequence of digits; that is, $a_n \in \{0, 1, 2, 3, 4, 5, 6, 7, 8, 9\}$. From Theorem 4.6.1 it follows that the series

$$\left\{ \sum_{k=1}^{n} \frac{a_k}{10^k} \right\}$$

is always convergent, because

$$0 \leq \frac{a_k}{10^k} \leq \frac{9}{10^k}$$

and

(1)
$$\sum_{k=1}^{\infty} \frac{9}{10^k} = \frac{9}{10} \left(1 + \frac{1}{10} + \frac{1}{10^2} + \cdots \right) = \frac{9}{10} \frac{1}{1 - \frac{1}{10}} = 1.$$

Therefore every sequence of digits corresponds to a number $A = \sum_{k=1}^{\infty} a_k / 10^k$ such that $0 \le A \le 1$.

In practice the number A is presented by writing the consecutive digits side by side and preceding them by a point and 0, for example 0.5772. Usually such a symbol does not describe the number entirely, because only a part of the sequence $\{a_n\}$ is taken into account. Instead we give a value close to A, namely,

$$A_n = \sum_{k=1}^{n} \frac{a_k}{10^k},$$

where n is the number of digits written after the point. In our example we have

$$0.5772 = \frac{5}{10} + \frac{7}{10^2} + \frac{7}{10^3} + \frac{2}{10^4}.$$

We can prove that in general

(2) $$A_n \le A \le A_n + \frac{1}{10^n}.$$

Indeed, since the sequence $\{A_n\}$ is nondecreasing, we have $A_n \le A$ for all n. In order to prove the second inequality we write

$$A = A_n + \sum_{k=n+1}^{\infty} \frac{a_k}{10^k}.$$

For the last sum we have

$$\sum_{k=n+1}^{\infty} \frac{a_k}{10^k} = \sum_{k=1}^{\infty} \frac{a_{n+k}}{10^{n+k}} \le \frac{1}{10^n} \sum_{k=1}^{\infty} \frac{9}{10^k} = \frac{1}{10^n},$$

which proves (2).

As another example let us consider the case when all the digits are 2. Then

$$A = \sum_{k=1}^{\infty} \frac{2}{10^k} = \frac{2}{10} \frac{1}{1 - \frac{1}{10}} = \frac{2}{9}.$$

For $n = 5$ we have

$$A_5 \le A \le A_5 + \frac{1}{10^5},$$

that is, $0.22222 \le \frac{2}{9} \le 0.22223$. We often write $\frac{2}{9} \approx 0.22222$. The number 0.22222 is the decimal approximation of $\frac{2}{9}$ up to 5 decimal places. A symbol such as 278.3023 is used to denote the number $278 + 0.3023$. The expression $A \approx 278.3023$ is read "A equals approximately 278.3023." In this book we adopt the convention that $A \approx 278.3023$ means that

$$278.3023 \le A < 278.3024.$$

In general, if we write $A \approx b_k \ldots b_0.a_1 \ldots a_n$, where $b_0, \ldots, b_k, a_1, \ldots, a_n$ are digits, then

$$b_k \ldots b_0.a_1 \ldots a_n \le A < b_k \ldots b_0.a_1 \ldots a_n + \frac{1}{10^n}.$$

At the beginning of this section we showed that every sequence of digits $0.a_1a_2a_3\ldots$ represents a number $0 \le x \le 1$. Now we will show that, conversely, every real number $0 \le x \le 1$ has such a representation. Since $0 = 0.000\ldots$ and $1 = 0.999\ldots$, we can assume that $0 < x < 1$. Let a_1 be the the greatest integer such that

$$\frac{a_1}{10} \le x.$$

Obviously, $a_1 \in \{0, 1, \ldots, 9\}$. Next, define a_2 to be the greatest integer such that

$$\frac{a_1}{10} + \frac{a_2}{10^2} \le x.$$

In general, define a_n to be the greatest integer such that

(3)
$$\frac{a_1}{10} + \frac{a_2}{10^2} + \cdots + \frac{a_n}{10^n} \le x.$$

Clearly, $a_n \in \{0, 1, \ldots, 9\}$ for all $n \in \mathbb{N}$. We will prove that $0.a_1a_2a_3\ldots$ is the decimal representation of x; that is,

(4)
$$x = \sum_{k=1}^{\infty} \frac{a_k}{10^k}.$$

By (3), we have $\sum_{k=1}^{\infty} a_k/10^k \le x$. Suppose that $\sum_{k=1}^{\infty} a_k/10^k < x$. Let n be the least integer such that

$$\frac{1}{10^n} \le x - \sum_{k=1}^{\infty} \frac{a_k}{10^k}$$

or equivalently

$$\sum_{k=1}^{\infty} \frac{a_k}{10^k} + \frac{1}{10^n} \le x.$$

This implies that

$$\frac{a_1}{10} + \frac{a_2}{10^2} + \cdots + \frac{a_{n-1}}{10^{n-1}} + \frac{a_n + 1}{10^n} \le x,$$

contradicting the definition of a_n. This proves (4).

Although, as we have proved, every $x \in [0, 1]$ has a decimal expansion, the decimal expansion is not unique. For example, $0.5 = 0.4999\ldots$ However, if we exclude sequences $a_1, a_2, a_3, \ldots$ for which there is an index n_0 such that $a_n = 9$ for all $n \geq n_0$, then every number can be represented by exactly one such a sequence of digits. To prove this, suppose that there are two different representations of a number x:

$$x = 0.a_1 a_2 a_3 \ldots, \qquad x = 0.b_1 b_2 b_3 \ldots$$

Let k be the least natural number for which $a_k \neq b_k$. We may assume that $a_k < b_k$. Let n be the least natural number greater than k such that $a_n \neq 9$ and let $c = a_n + 1$. Then

$$x = 0.a_1 a_2 a_3 \ldots < 0.a_1 a_2 \ldots a_{n-1} c < 0.b_1 b_2 \ldots b_n \leq 0.b_1 b_2 b_3 \ldots = x.$$

This contradiction shows that x cannot have two such representations.

Notice that, since any real number x can be written $x = a + b$ where a is an integer and $b \in [0, 1)$, x has a decimal expansion, which is unique under the assumption of exclusion discussed above.

4.9 POWER SERIES

In Section 4.4 we proved that the geometric series

$$(1) \qquad \left\{ \sum_{k=0}^{n} x^k \right\}$$

converges for every x in $(-1, 1)$. The sum depends on x, and thus it represents a function defined on the interval $(-1, 1)$. We have proved that the function is $1/(1 - x)$. The geometric series is an example of a power series.

Another example of a power series is

$$(2) \qquad \left\{ \sum_{k=1}^{n} \frac{(-1)^{k+1}}{k} x^k \right\}.$$

This series is also convergent for every x in $(-1, 1)$, because $|(-1)^{k+1} x^k / k| \leq |x|^k$ for all $k \in \mathbb{N}$ and the series $\{\sum_{k=1}^{n} |x|^k\}$ converges for every $x \in (-1, 1)$.

The name *power series* is traditionally given to any series of the form

$$(3) \qquad \left\{ \sum_{k=0}^{n} a_k x^k \right\}$$

where the coefficients $a_0, a_1, a_2, \ldots$ are constants.

Power series play a very important role in mathematical analysis. They allow us to introduce several new functions of fundamental significance. For example, series (2) is connected with the logarithm function.

One of the basic problems concerning power series is determining the set of all points x for which a given power series converges. Obviously, every power series converges for $x = 0$, and the sum is a_0.

—The first term of the series $\{\sum_{k=0}^{n} a_k x^k\}$ is $a_0 x^0$. In particular, if $x = 0$, we get $a_0 0^0$. What is the value of 0^0?

—1.

—Is this something obvious?

—Not at all. On the one hand, we have $\lim_{x \to 0+} x^0 = 1$ and thus it is reasonable to define $0^0 = 1$. On the other hand, however, $\lim_{x \to 0+} 0^x = 0$. Finally, using methods discussed in Chapter 5, one can show that $\lim_{x \to 0+} x^x = 1$.

—Oh! I see, since we get 1 more often than 0, we define $0^0 = 1$.

—Not really. Here, defining $0^0 = 1$ is a matter of convenience. We want the first term of the series $\{\sum_{k=0}^{n} a_k x^k\}$ to be the constant a_0. Note that all the terms have the same form because $0^0 = 1$. For the same reason we define $0! = 1$. In this case too, there are other reasons for this choice.

4.9.1 Theorem   *If a power series*

$$(4) \qquad \left\{ \sum_{k=0}^{n} a_k x^k \right\}$$

is convergent at some point $x_0 \neq 0$, then there exists a number $r > 0$ such that the series is absolutely convergent for all $x \in (-r, r)$ and divergent (that is, not convergent) for $|x| > r$. The case $r = \infty$ is also admitted, which means that the series converges for all real x.

Let us recall that the absolute convergence of (4) implies that the series

$$(5) \qquad \left\{ \sum_{k=0}^{n} |a_k||x|^k \right\}$$

is convergent. By Theorem 4.6.3 every absolutely convergent series is convergent. Thus if we erase the word "absolutely" in Theorem 4.9.1 the theorem remains true but becomes weaker.

Proof of 4.9.1.   Assume that series (4) is convergent for some x_0. Set $z = |x_0|$. Then $\lim_{n \to \infty} a_n z^n = 0$, by Theorem 4.5.1. Hence $|a_n z^n| < M$ for all $n = 0, 1, 2, \ldots$ and some M. If $|x| < z$, then

$$|a_n||x|^n = |a_n z^n| \frac{|x|^n}{z^n} \leq M \frac{|x|^n}{z^n}$$

for $n = 0, 1, 2, \ldots$. Since $|x|/z < 1$, the series $\{\sum_{k=1}^{n} M|x|^k/z^k\}$ is convergent, and hence series (5) is convergent. This means that series (4) is absolutely convergent for all $|x| < z$.

If the set of all x for which the series (5) converges is bounded, then the least upper bound of that set will be denoted by r. If $x \in (-r, r)$, then there exists a number z, $|x| < z < r$, such that the series

$$\left\{ \sum_{k=0}^{n} a_k z^k \right\}$$

converges. This implies, by the comparison test, convergence of (5). We have thus proved the absolute convergence of (4) in $(-r, r)$. If $r = \infty$ the proof is the same, only the restriction $z < r$ drops out.

It remains to prove that series (4) diverges for $|x| > r$, but this follows from the definition of the least upper bound r and the first part of this proof.

The number r in the above theorem is called the *radius of convergence* of series (4). Although ∞ is not a number, it is convenient to say "the radius is infinity" and write $r = \infty$. The set of all numbers x for which the series (4) converges is usually called the *interval of convergence* of that series. Theorem 4.9.1 implies the following:

If $r = 0$, then the interval of convergence is reduced to the single point $\{0\}$.

If $r = \infty$, then the interval of convergence is the entire real line.

If $0 < r < \infty$, then the interval of convergence takes one of the following forms:

$$(-r, r), \ [-r, r), \ (-r, r], \ \text{or} \ [-r, r].$$

Note that Theorem 4.9.1 says nothing about the convergence of the series at the endpoints. Those have to be checked separately. For example, the interval of convergence of series (1) is $(-1, 1)$, because both series

$$\left\{ \sum_{k=0}^{n} 1^k \right\} \quad \text{and} \quad \left\{ \sum_{k=0}^{n} (-1)^k \right\}$$

diverge. The interval of convergence of series (2) is $(-1, 1]$. In fact, in Section 4.36 we will show that the series

$$\left\{ \sum_{k=1}^{n} \frac{(-1)^{k+1}}{k} 1^k \right\}$$

converges (see Theorem 4.36.1). Divergence of the series

$$\left\{ \sum_{k=1}^{n} \frac{(-1)^{k+1}}{k} (-1)^k \right\}$$

follows easily from Theorem 7.20.1.

In applications of power series discussed in the following sections convergence at the endpoints of the interval of convergence will not play any role. Therefore, for the sake of simplicity, by the interval of convergence we will always mean the open interval $(-r, r)$.

4.9.2 Theorem *If $0 < q < 1$, then $\lim_{n \to \infty} nq^n = 0$.*

Proof. We have $\lim_{n \to \infty} (n + 1)/nq = q < 1$. Thus there exists a number n_0 such that $(n + 1)/nq < 1$ for all $n > n_0$. Hence $(n+1)q^{n+1} < nq^n$ for $n > n_0$. Thus the sequence $\{nq^n\}$ is decreasing for $n > n_0$. Since it is also bounded, it converges to some limit k. From the equality $(n + 1)q^{n+1} = qnq^n + q^{n+1}$ we obtain, by letting $n \to \infty$, $k = qk$. Since $q \neq 1$, we conclude that $k = 0$, proving the theorem.

From the above theorem it follows that if $|x| > 1$, then $\lim_{n \to \infty} n/|x|^n = 0$, and hence $\lim_{n \to \infty} |x|^n/n = \infty$. Thus, by Theorem 4.5.1, for $|x| > 1$ the series (2) is not convergent.

Exercises 4.9

1. Complete the proof of Theorem 4.9.1 by showing that series (4) diverges when $|x| > r$.

2. Prove that if the radius of convergence of $\{\sum_{k=0}^{n} a_k x^k\}$ is r and $m \in \mathbb{N}$, then the radius of convergence of $\{\sum_{k=0}^{n} a_k x^{km}\}$ is $r^{1/m}$.

3. Prove that if $0 < q < 1$ and $k \in \mathbb{N}$, then $\lim_{n \to \infty} n^k q^n = 0$.

4.10 SEQUENCES OF FUNCTIONS

The partial sums $s_n(x) = \sum_{k=0}^{n} a_k x^k$ form a sequence of functions of x. Generally, by a sequence of functions we mean a sequence $\{f_n\}$ whose terms are functions defined on a common interval I. A sequence of functions $\{f_n\}$ becomes an ordinary (numerical) sequence, if all its terms are evaluated at an arbitrary point in I. Thus a sequence of functions $\{f_n\}$ associates with each point $x \in I$ a numerical sequence $\{f_n(x)\}$.

If, for each point $x \in I$, the sequence $\{f_n(x)\}$ converges to a limit $f(x)$, then the limits define a function on I. In such a case we say that the sequence $\{f_n\}$ converges or is convergent to f and write

$$\lim_{n \to \infty} f_n(x) = f(x) \text{ for } x \in I.$$

This type of convergence is called *pointwise convergence*.

Pointwise convergence is a weak convergence. The famous Cauchy believed that the pointwise limit of a sequence of continuous functions is continuous. He was wrong, which is easily seen from the following example.

For $n = 1, 2, 3, \ldots$, define, as shown in Fig. 4.2,

$$f_n(x) = \begin{cases} 1 & \text{for } x > \dfrac{1}{n} \\ nx & \text{for } 0 \leq x \leq \dfrac{1}{n} \\ 0 & \text{for } x < 0 \end{cases}$$

The limit of this sequence of functions is a discontinuous function:

$$f(x) = \begin{cases} 1 & \text{for } x > 0 \\ 0 & \text{for } x \leq 0. \end{cases}$$

To ensure the continuity of the limit of a sequence of continuous functions a stronger type of convergence is needed.

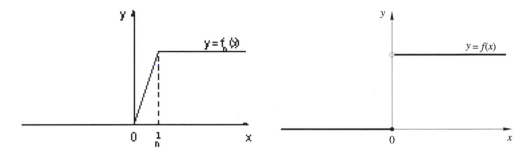

Figure 4.2 A sequence of continuous functions with a discontinuous limit.

Exercises 4.10

1. Find the limit of the sequence of functions $f_n(x) = x^n$ for $x \in [0, 1]$.

2. Find the limits for $x \in \mathbb{R}$:

   (a) $\lim_{n \to \infty} \dfrac{2x + nx^2}{1 + nx^2}$.

   (b) $\lim_{n \to \infty} \dfrac{nx + x^2 + nx^3}{1 + nx^2}$.

   (c) $\lim_{n \to \infty} \dfrac{(n + 1)x + n^2x^3}{1 + n^2x^2}$.

3. Prove that the sequence of functions $\sqrt{x},\ \sqrt{x + \sqrt{x}},\ \sqrt{x + \sqrt{x + \sqrt{x}}}, \ldots$ converges on $[0, \infty)$. Find the limit.

4.11 UNIFORM CONVERGENCE

We adopt the following definition:

4.11.1 Definition We say that a sequence of functions $\{f_n\}$ is *uniformly convergent* to a function f on an interval I, if there exists a sequence of positive numbers $\{\varepsilon_n\}$ such that $\lim_{n \to \infty} \varepsilon_n = 0$ and the inequality $|f_n(x) - f(x)| < \varepsilon_n$ holds for all sufficiently large $n \in \mathbb{N}$ and all $x \in I$. We then write $f_n \rightrightarrows f$ in I.

In view of Theorem 4.1.3 we then have $\lim_{n \to \infty}(f_n(x) - f(x)) = 0$; that is, $\lim_{n \to \infty} f_n(x) = f(x)$ for every $x \in I$. Thus *every uniformly convergent sequence of functions converges (pointwise) to the same limit.*

The sequence of functions $f_n(x) = x^n$ is convergent to 0 uniformly on the interval $[0, \alpha]$ for any $0 < \alpha < 1$, because then $|x^n - 0| < \alpha^n$ in $[0, \alpha]$ and $\lim_{n \to \infty} \alpha^n = 0$. Although we have $\lim_{n \to \infty} x^n = 0$ for every $x \in [0, 1)$, the sequence $\{x^n\}$ is not uniformly convergent on $[0, 1)$, because, for all $n \in \mathbb{N}$, the least ε_n for which $|x^n - 0| < \varepsilon_n$ holds in the whole interval $[0, 1)$ is $\varepsilon_n = 1$.

4.11.2 Theorem *If $f_n \rightrightarrows f$ and $g_n \rightrightarrows g$ on I, then $f_n + g_n \rightrightarrows f + g$ and $f_n - g_n \rightrightarrows f - g$ on I.*

Proof. By assumption there exist two sequences of positive numbers $\{\varepsilon_n\}$ and $\{\delta_n\}$ tending to zero such that $|f_n - f| < \varepsilon_n$ and $|g_n - g| < \delta_n$ for all sufficiently large $n \in \mathbb{N}$. Hence

$$|(f_n + g_n) - (f + g)| < \varepsilon_n + \delta_n \quad \text{and} \quad |(f_n - g_n) - (f - g)| < \varepsilon_n + \delta_n.$$

This proves the theorem, for the sequence $\{\varepsilon_n + \delta_n\}$ converges to zero.

Analogously one can, taking some precautions, prove similar theorems for multiplication and division. However, those theorems will not be needed here.

A constant sequence of constant functions $f_n(x) = c$ is uniformly convergent, because for every sequence of positive numbers $\{\varepsilon_n\}$ we have $|f_n(x) - c| = 0 < \varepsilon_n$. This implies, by Theorem 4.11.2, that if a sequence $\{g_n\}$ is uniformly convergent, then the sequence $\{g_n + c\}$ is also uniformly convergent.

The property of uniform convergence presented in the following theorem is the main reason for which this type of convergence is important.

4.11.3 Theorem *If a sequence of continuous functions $\{f_n\}$ is uniformly convergent in an interval I, then its limit is a continuous function on I.*

Proof. Let $\lim_{n \to \infty} f_n(x) = f(x)$. From the equality

$$f(x) - f(x_0) = f(x) - f_n(x) + f_n(x) - f_n(x_0) + f_n(x_0) - f(x_0)$$

we get

$$|f(x) - f(x_0)| \le |f(x) - f_n(x)| + |f_n(x) - f_n(x_0)| + |f_n(x_0) - f(x_0)|.$$

Because of the uniform convergence of $\{f_n\}$ we have, for any $x, x_0 \in I$ and sufficiently large $n \in \mathbb{N}$,

$$|f(x) - f(x_0)| < \varepsilon_n + |f_n(x) - f_n(x_0)| + \varepsilon_n$$

where the ε_n form a sequence of positive numbers convergent to 0. For each $\varepsilon > 0$ there exists an index m such that $\varepsilon_m < \frac{1}{3}\varepsilon$ and thus

$$|f(x) - f(x_0)| \le \frac{2}{3}\varepsilon + |f_m(x) - f_m(x_0)|.$$

Since the function f_m is continuous, there exists a number $\delta > 0$ such that

$$|f_m(x) - f_m(x_0)| < \frac{1}{3}\varepsilon$$

for all $x \in (x_0 - \delta, x_0 + \delta)$. Hence

$$|f(x) - f(x_0)| < \varepsilon \quad \text{for} \quad |x - x_0| < \delta.$$

This proves that f is continuous at x_0, and since x_0 is an arbitrary point of the interval I, this proves the continuity of f on I.

Exercises 4.11

1. Prove that $f_n \rightrightarrows f$ on I if and only if for every $\varepsilon > 0$ there exists $n_0 \in \mathbb{N}$ such that for all $n \ge n_0$ and all $x \in I$ we have $|f_n(x) - f(x)| < \varepsilon$.

2. Let $f, f_1, f_2, \ldots$ be continuous functions on a bounded closed interval $[a, b]$. Prove that $f_n \rightrightarrows f$ on $[a, b]$ if and only if

$$\lim_{n \to \infty} \max_{x \in [a, b]} |f_n(x) - f(x)| = 0.$$

3. Discuss the uniform convergence of the following sequences of functions:

(a) $\left\{ \dfrac{2x + nx^2}{1 + nx^2} \right\}.$ (b) $\left\{ \dfrac{nx + x^2 + nx^3}{1 + nx^2} \right\}.$ (c) $\left\{ \dfrac{(n + 1)x + n^2x^3}{1 + n^2x^2} \right\}.$

4. True or false? If $f_n \rightrightarrows f$ and $g_n \rightrightarrows g$ in $\mathbb{R}$, then $f_n g_n \rightrightarrows f g$.

5. True or false? Let f, f_n, g, g_n be continuous functions on a bounded closed interval $[a, b]$. If $f_n \rightrightarrows f$ and $g_n \rightrightarrows g$ in $[a, b]$, then $f_n g_n \rightrightarrows f g$ on $[a, b]$.

6. Formulate and prove a theorem on the uniform convergence of sequences of the form $\{f_n / g_n\}$.

7. Let f be a function with bounded derivative on $\mathbb{R}$ and let $f_n(x) = f(x + \frac{1}{n})$ for $n \in \mathbb{N}$. Prove that $f_n \rightrightarrows f$ on $\mathbb{R}$.

8. Let $\{f_n\}$ be a uniformly convergent sequence of continuous functions on $[a, b]$ and let $c \in [a, b]$. Prove that $\lim_{n \to \infty} \lim_{x \to c} f_n(x) = \lim_{x \to c} \lim_{n \to \infty} f_n(x)$.

4.12 UNIFORM CONVERGENCE OF POWER SERIES

A power series is said to be uniformly convergent on an interval, if the sequence of partial sums is uniformly convergent on that interval.

4.12.1 Theorem   *If r is the radius of convergence of a power series, then for every $0 < b < r$ the series is uniformly convergent on $[-b, b]$.*

Proof.   Let r be the radius of convergence of the power series

(1)
$$\left\{ \sum_{k=0}^{n} a_k x^k \right\}.$$

If $0 < b < c < r$, then the series

(2)
$$\left\{ \sum_{k=0}^{n} |a_k| c^k \right\}$$

converges. By comparison with (2), the series $\{\sum_{k=0}^{n} a_k x^k\}$ converges for every $x \in [-b, b]$. Let $s(x) = \sum_{k=0}^{\infty} a_k x^k$, for $x \in [-b, b]$. If we denote $t_n = \sum_{k=0}^{n} |a_k| c^k$ and $t = \sum_{k=0}^{\infty} |a_k| c^k$, then we have

$$\left| s(x) - \sum_{k=0}^{n} a_k x^k \right| = \left| \sum_{k=n+1}^{\infty} a_k x^k \right| \leq \sum_{k=n+1}^{\infty} |a_k x^k| \leq \sum_{k=n+1}^{\infty} |a_k| c^k = t - t_n$$

for every $x \in [-b, b]$. Since $\lim_{n \to \infty} t_n = t$, this proves the uniform convergence of (1).

4.12.2 Theorem   *The function $s(x) = \sum_{k=0}^{\infty} a_k x^k$ is continuous in the interval of convergence of the power series $\{\sum_{k=0}^{n} a_k x^k\}$.*

Proof.   If x is an arbitrary point in the interval of convergence $(-r, r)$ of a power series, then there exists an interval $[-b, b]$, $0 < b < r$, containing x. By Theorem 4.12.1 the series converges uniformly and, by Theorem 4.11.3, the sum is a continuous function on $(-b, b)$. This proves that the sum is continuous at x, which is an arbitrary point of $(-r, r)$. Thus the sum is a continuous function on the entire interval of convergence $(-r, r)$.

Exercises 4.12

1. Show that the series $\{\sum_{k=0}^{n} x^k\}$ is not uniformly convergent on $(-1, 1)$.

2. A series of functions $\{\sum_{k=1}^{n} f_k(x)\}$ is said to converge uniformly on an interval I if the sequence of partial sums $s_n(x) = f_1(x) + \cdots + f_n(x)$ converges uniformly on I. Prove the following theorem, known as the *Weierstrass Test:*
    If $|f_n(x)| \le M_n$ for all $n \in \mathbb{N}$ and every $x \in I$, and the series $\{\sum_{k=1}^{n} M_k\}$ converges, then the series $\{\sum_{k=1}^{n} f_k(x)\}$ converges uniformly on I.

3. Find an example of a sequence of continuous functions f_n on $[0, 1]$ such that the series $\{\sum_{k=1}^{n} f_k(x)\}$ converges uniformly on $[0, 1]$ but the series $\{\sum_{k=1}^{n} \max_{[0,1]} |f_k(x)|\}$ diverges. Is it a counterexample to the Weierstrass Test?

4. Prove that if $|f_n(x)| \le g_n(x)$ for all $n \in \mathbb{N}$ and every $x \in [a, b]$, and the series $\{\sum_{k=1}^{n} g_k(x)\}$ converges uniformly on $[a, b]$, then $\{\sum_{k=1}^{n} f_k(x)\}$ converges uniformly in $[a, b]$.

4.13 DIFFERENTIATION OF SEQUENCES OF FUNCTIONS

We shall prove the following theorem.

4.13.1 Theorem   *If $f_n \rightrightarrows f$ and $f_n' \rightrightarrows g$ on an interval I and the functions f_n' are continuous, then f is differentiable and $f' = g$.*

This theorem will be used to prove a theorem on differentiation of power series and therefore it is stated in the above unpretentious form. Its formulation can be strengthened in various directions, but then the proofs become more difficult.

Proof.   The uniform convergence of $\{f_n'\}$ means that there exists a sequence of positive numbers ε_n convergent to 0 such that

$$|f_n'(x) - g(x)| < \varepsilon_n$$

for all $x \in (a, b)$ and for all sufficiently large $n \in \mathbb{N}$. Since the functions f_n' are continuous, so is the limit g, by Theorem 4.11.3. Thus, if $x_0 \in (a, b)$, then for every $\varepsilon > 0$ there exists a number $\delta > 0$ such that

$$|g(x) - g(x_0)| < \varepsilon \quad \text{for} \quad |x - x_0| < \delta.$$

By the mean value theorem (Theorem 3.20.1) we have

(1)
$$\left| \frac{f_n(x) - f_n(x_0)}{x - x_0} - g(x_0) \right| = |f_n'(\xi) - g(x_0)|$$

$$\leq |f_n'(\xi) - g(\xi)| + |g(\xi) - g(x_0)| \leq \varepsilon_n + \varepsilon$$

for $|x - x_0| < \delta$, where ξ is a number between x and x_0 and hence satisfies the inequality $|\xi - x_0| < \delta$. By letting $n \to \infty$ in (1) we obtain

$$\left| \frac{f(x) - f(x_0)}{x - x_0} - g(x_0) \right| \leq \varepsilon \quad \text{for} \quad |x - x_0| < \delta.$$

This proves that

$$\lim_{x \to x_0} \frac{f(x) - f(x_0)}{x - x_0} = g(x_0);$$

that is, $f'(x_0) = g(x_0)$. Since the point x_0 can be chosen arbitrarily in (a, b), the proof is finished.

The following theorem is of critical importance in mathematical analysis.

4.13.2 Theorem *The power series*

(2)
$$\left\{ \sum_{k=0}^{n} a_k x^k \right\}$$

and

(3)
$$\left\{ \sum_{k=1}^{n} k a_k x^{k-1} \right\}$$

have the same radius of convergence. Moreover,

$$\left(\sum_{k=0}^{\infty} a_k x^k \right)' = \sum_{k=1}^{\infty} k a_k x^{k-1}$$

in the interval of convergence.

Note that this theorem says that a power series can be differentiated term by term.

Proof. Let r be the radius of convergence of (2) and let x be an arbitrary fixed number such that $0 < |x| < r$. Choose b so that $|x| < b < r$. Since the sequence $\{n |x/b|^n\}$ converges to 0, it is bounded. There is thus a number M such that

$$|n a_n x^{n-1}| = \left| \frac{1}{x} n \left(\frac{x}{b} \right)^n a_n b^n \right| \leq \frac{1}{|x|} \left(n \left| \frac{x}{b} \right|^n \right) |a_n| b^n \leq M |a_n| b^n.$$

Because the series $\{\sum_{k=1}^{n}|a_k|b^k\}$ converges, series (3) converges at x. Thus (3) converges at every point at which (2) converges.

Conversely, series (2) is convergent at every point at which (3) is convergent, for $|a_n x^n| \le |x|n|a_n||x^{n-1}|$. Therefore series (2) and (3) have the same radius of convergence.

Now let

$$s_n(x) = \sum_{k=0}^{n} a_k x^k \quad \text{and} \quad t_n(x) = \sum_{k=1}^{n} k a_k x^{k-1}.$$

Note that, for all $n \in \mathbb{N}$, s_n and t_n are polynomials and we have $s_n' = t_n$. The sequences $\{s_n\}$ and $\{t_n\}$ are uniformly convergent on each interval $[-b, b]$ such that $0 < b < r$ and hence $f' = g$ in $[-b, b]$. This implies the equality over the entire interval of convergence $(-r, r)$.

Exercises 4.13

1. Give an example of a sequence $\{f_n\}$ of differentiable functions on an interval I such that $f_n \rightrightarrows 0$ on I, but the sequence $\{f_n'\}$ is not convergent on I.

2. Prove that if both series $\{\sum_{k=1}^{n} f_k(x)\}$ and $\{\sum_{k=1}^{n} f_k'(x)\}$ converge uniformly on I and the f_k' are continuous, then the function $f(x) = \sum_{k=1}^{\infty} f_k(x)$ is differentiable on I and $f'(x) = \sum_{k=1}^{\infty} f_k'(x)$.

3. Find the sums:

    (a) $\displaystyle\sum_{k=1}^{\infty} k x^k$, $|x| < 1$,   (b) $\displaystyle\sum_{k=1}^{\infty} k^2 x^k$, $|x| < 1$.

4. Find the sums:

    (a) $\displaystyle\sum_{k=1}^{\infty} \frac{k}{3^k}$,   (b) $\displaystyle\sum_{k=1}^{\infty} \frac{k^2}{7^k}$.

4.14 EXPONENTIAL FUNCTION

We now approach the didactic problem of introducing the exponential function a^x. This function is perhaps the most important function in calculus and is introduced in a great many ways. For instance it can be introduced as a generalization of the power a^n to nonnatural values of the exponent n. First we introduce exponents $-n$ by assuming that $a^{-n} = 1/a^n$. Next we define $a^{1/n}$ as the only solution of the equation $x^n = a$.

—But there are equations that have no solution, for example $x^2 = -1$.

—Yes, and for that reason we assume that $a > 0$. We have to remember that the exponential function is an extension of the power with positive base.

—How do we know that a solution exists for $a > 0$ and that it is unique?

—This will be shown in Section 4.19.

—And how do we define a^x for other values of x?

—The exponents m/n are introduced by defining $a^{m/n} = \left(a^{1/n}\right)^m$. Finally, if x is an irrational number we approximate it by a sequence of rational numbers x_n and then we prove that the sequence $\{a^{x_n}\}$ is convergent. The value of a^x is defined as the limit of $\{a^{x_n}\}$.

—Since there are infinitely many sequences of rational numbers convergent to x, we may obtain infinitely many values for a^x.

—No. One can prove that the limit does not depend on the choice of the sequence approximating x.

—This way of introducing the exponential function seems to be long and complicated.

—So it is. For this reason authors often introduce the exponential function e^x using the power series

$$e^x = 1 + \frac{x}{1!} + \frac{x^2}{2!} + \frac{x^3}{3!} + \cdots.$$

—What is the meaning of e?

—This letter has, at the beginning, no separate meaning. The symbol e^x has to be regarded as a whole, like $f(x)$.

—And what is the connection with a^x?

—Using the function e^x one can define a^x by a proper construction.

—This way looks artificial.

—Because of this there were mathematicians who were looking for another, possibly simpler definition. For example, the English mathematician Godfrey Harold Hardy (1877–1947) used the notion of integral. One can also use the functional equation $f(x + y) = f(x)f(y)$ as the starting point.

We see that there are many possible ways of introducing the exponential function. It is difficult to decide which way is the best. In this book we shall start from the fact that the derivative of e^x is e^x. This approach seems to be the most useful one.

4.14.1 Definition   The *exponential function e^x* is the only solution of the differential equation

(1) $$y' = y$$

satisfying the initial condition $y(0) = 1$.

In other words, the exponential function e^x is a function that is invariant under differentiation and has the value 1 at 0. In order to be able to use this definition we have to answer two questions: Does a solution exist? And is it unique?

4.15 EXISTENCE OF A SOLUTION OF THE EQUATION $y' = y$

The existence of a solution of the equation

(1) $$y' = y, \qquad y(0) = 1$$

follows immediately as a very particular case of a general theorem on the existence of solutions of differential equations. However, we can omit this general theory by basing the proof on a power series argument.

Let us assume for a while that there exists a solution of (1) that can be represented by a power series:

$$(2) \qquad\qquad y(x) = \sum_{k=0}^{\infty} a_k x^k.$$

Then by Theorem 4.13.2 we have

$$y'(x) = \sum_{k=0}^{\infty} (k+1)a_{k+1} x^k.$$

If we assume that

$$(3) \qquad\qquad a_{k+1} = \frac{1}{k+1} a_k,$$

then equation (1) is satisfied. In particular, from (3) we get

$$a_1 = \frac{a_0}{1}, \qquad a_2 = \frac{a_0}{1 \cdot 2}, \qquad a_3 = \frac{a_0}{1 \cdot 2 \cdot 3},$$

and in general

$$a_k = \frac{a_0}{k!},$$

where $k! = 1 \cdot 2 \cdots k$. Since we wish to have $y(0) = 1$, we have to take $a_0 = 1$. We have thus proved that the function

$$(4) \qquad\qquad e^x = 1 + \frac{x}{1!} + \frac{x^2}{2!} + \cdots$$

satisfies (1) in the interval of convergence of the power series. We will prove that the series converges for every x, which means the the interval of convergence is the entire real line.

Let x be an arbitrary fixed number and let m be a natural number greater than $|x|$. We have

$$\frac{|x|^{m+1}}{(m+1)!} = \frac{|x|^m}{m!} \cdot \frac{|x|}{m+1} < \frac{|x|^m}{m!} \cdot \frac{|x|}{m}$$

and by induction

$$(5) \qquad\qquad \frac{|x|^{m+k}}{(m+k)!} < \frac{|x|^m}{m!} \cdot \left(\frac{|x|}{m}\right)^k.$$

Since, for every $n \in \mathbb{N}$, we have

$$\sum_{k=1}^{n} \frac{|x|^m}{m!}\left(\frac{|x|}{m}\right)^k = \frac{|x|^m}{m!} \sum_{k=1}^{n}\left(\frac{|x|}{m}\right)^k$$

and the series

$$\left\{ \sum_{k=1}^{n}\left(\frac{|x|}{m}\right)^k \right\}$$

is convergent, convergence of the series

$$\left\{ \sum_{k=1}^{n} \frac{|x|^{m+k}}{(m+k)!} \right\}$$

follows by (5). This implies convergence of the series in (4) for every $x \in \mathbb{R}$.

—We assumed that there was a power series satisfying (1). How can we assume this when our aim was just to prove existence of a solution?

—There is no vicious circle in this argument because the assumption of existence served only to find the coefficients. Having found those coefficients, we disregard our assumption and we prove that the series is convergent and represents a function with the desired properties.

—Would it not be simpler to just start from series (4) since we have to use it anyway?

—Many authors do so. However, such a definition is a little artificial because there is no explanation of the choice of coefficients. The actual explanation is the differential equation (1). We could say that e^x is the only function expandable into a power series and satisfying (1).

—But in Definition 4.14.1 there is no assumption that the function is expandable. Is that definition correct?

—Yes, it is. One can prove that each function satisfying (1) is expandable, and this ought to be proved in order to show that the definition is correct. To this aim we shall need a theorem on composite functions.

Exercise 4.15

1. Use the method presented in this section to prove that the equation $y' = -y$, $y(0) = 1$, has a solution defined for all $x \in \mathbb{R}$.

4.16 DIFFERENTIATION OF COMPOSITE FUNCTIONS

Given two functions f and g the function defined by

$$(1) \qquad\qquad F(x) = f(g(x))$$

is called a composite function. In Section 3.12 we proved that the composition of two continuous functions is a continuous function. The following theorem shows that the same is true for differentiability. Formula (2) is often called the *chain rule*.

4.16.1 Theorem   *If a function g is differentiable at a point x_0 and a function f is differentiable at the point $g(x_0)$, then the composite function $F = f \circ g$ is differentiable at x_0 and*

$$(2) \qquad\qquad F'(x_0) = f'(g(x_0))g'(x_0).$$

Proof.   Set $t_0 = g(x_0)$. Let f be defined in some interval (a, b) containing t_0. We introduce an auxiliary function ρ whose value at $t \neq t_0$ is given by

$$(3) \qquad\qquad \rho(t) = \frac{f(t) - f(t_0)}{t - t_0} - f'(t_0).$$

and whose value at t_0 is $\rho(t_0) = 0$. Then the function ρ is defined on the entire interval (a, b) and continuous at t_0. From (3) we get for $t \neq t_0$

$$(4) \qquad f(t) - f(t_0) = (t - t_0)(f'(t_0) + \rho(t)).$$

Since this equality holds also for $t = t_0$, it holds on the entire interval (a, b).

The function g, being differentiable at x_0, is continuous at that point and thus there exists a number $\delta > 0$ such that $g(x) \in (a, b)$ whenever $|x - x_0| < \delta$. In view of (4), for $x \neq x_0$, we can write

$$\frac{F(x) - F(x_0)}{x - x_0} = \frac{f(g(x)) - f(g(x_0))}{x - x_0} = \frac{g(x) - g(x_0)}{x - x_0}\Big(f'(g(x_0)) + \rho(g(x))\Big)$$

Hence, by taking the limit as $x \to x_0$, we obtain $F'(x_0) = g'(x_0)f'(g(x_0))$, proving the theorem.

Exercises 4.16

1. Find the derivatives of the following functions. Assume that f, g, and h are differentiable.
    (a) $f \circ (g + h)$,   (b) $f \circ (gh)$,   (c) $f \circ f \circ f$.
2. Decide whether the following are possible:
    (a) f is differentiable, and g is not differentiable, and $f \circ g$ is differentiable.
    (b) f is differentiable, and g is not differentiable, and $g \circ f$ is differentiable.
    (c) f and g are not differentiable, and $f \circ g$ is differentiable.

4.17 UNIQUENESS OF THE SOLUTION OF THE EQUATION $y' = y$

We have defined the exponential function as the unique solution of the differential equation $y' = y$, $y(0) = 1$. In Section 4.15 we proved the existence of such a solution. It is easy to show that the solution found is unique. Indeed, assume that there are two solutions f and g and consider the auxiliary function

$$(1) \qquad h(x) = f(x)g(x_0 - x),$$

where x_0 is an arbitrary fixed point. Then

$$h'(x) = f'(x)g(x_0 - x) - f(x)g'(x_0 - x).$$

To understand the calculations one has to clearly distinguish the function $g'(x_0 - x)$ from the function $(g(x_0 - x))' = -g'(x_0 - x)$.

Since f and g are solutions of $y' = y$, we have

$$h'(x) = f(x)g(x_0 - x) - f(x)g(x_0 - x) = 0.$$

Thus h is a constant function. Consequently we have $h(x_0) = h(0)$, which means

$$f(x_0)g(0) = f(0)g(x_0),$$

or, in view of the condition $y(0) = 1$,

$$f(x_0) = g(x_0).$$

Since x_0 is an arbitrary number, f and g are identical and the uniqueness is proved.

Exercise 4.17

1. Prove the uniqueness of solution of the equation $y' = -y$, $y(0) = 1$.

4.18 PROPERTIES OF THE EXPONENTIAL FUNCTION

The function e^x is the only function equal to its derivative that has the value 1 at 0. Since it is differentiable it is of course continuous (this also follows from the fact that it can be expanded in a power series). In the previous section we proved that the function

$$h(x) = f(x)g(x_0 - x) = e^x e^{x_0 - x}$$

is constant. Hence

$$e^0 e^{x_0} = e^x e^{x_0 - x}.$$

Substituting $x_0 = x + y$ we get (since $e^0 = 1$),

(1) $$e^x e^y = e^{x+y}.$$

From the expansion

(2) $$e^x = 1 + \frac{x}{1!} + \frac{x^2}{2!} + \frac{x^3}{3!} + \cdots$$

it is easily seen that the function e^x is positive for $x \geq 0$. By (1) one can easily show that it is positive also for $x < 0$. In fact, for $y = -x$ we have $e^x e^{-x} = e^0 = 1$, and hence

$$e^{-x} = \frac{1}{e^x}.$$

Thus, the exponential function is positive for all $x \in \mathbb{R}$. It is also an increasing function, since its derivative is positive.

We are going to prove that e^x assumes each positive value exactly once. Since we may write

$$e^x = 1 + x + \sum_{k=2}^{\infty} \frac{x^k}{k!},$$

for every positive x we have

(3) $$1 + x < e^x$$

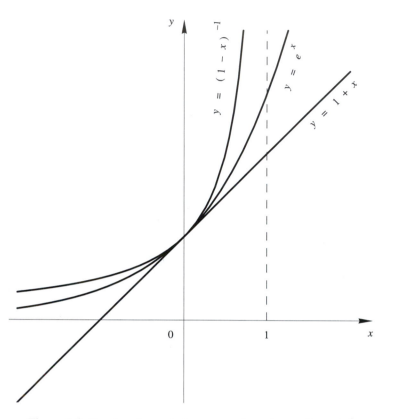

Figure 4.3 Graphs of $y = 1 + x$, $y = e^x$, and $y = (1 - x)^{-1}$.

(see Fig. 4.3). Let a be any positive number. Taking in turn a and $1/a$ we get

$$a < 1 + a < e^a \quad \text{and} \quad \frac{1}{a} < 1 + \frac{1}{a} < e^{1/a},$$

and hence

$$e^{-1/a} < a < e^a.$$

Since e^x is continuous it assumes each value between $e^{-1/a}$ and e^a, and in particular it assumes the value a. Moreover e^x assumes a only once, because it is an increasing function.

Exercises 4.18

1. Prove that inequality (3) holds also for $x < 0$.
2. Prove that for $x < 1$ we have $e^x \leq (1 - x)^{-1}$ (see Figure 4.3).

4.19 ARBITRARY EXPONENTIAL FUNCTIONS

We have proved that for every $a > 0$ there is a unique number c such that

$$a = e^c.$$

That number is called the *natural logarithm* of a and is denoted by

$$c = \ln a.$$

The two equalities are equivalent; thus we have

(1) $$a = e^{\ln a} \quad \text{for} \quad a > 0.$$

In view of

(2) $$e^x e^y = e^{x+y}$$

we have also

$$a^2 = e^{\ln a} e^{\ln a} = e^{2\ln a},$$
$$a^3 = e^{2\ln a} e^{\ln a} = e^{3\ln a},$$

and generally

(3) $$a^n = e^{n\ln a} \quad \text{for} \quad n \in \mathbb{N}.$$

Moreover we have

$$a^0 = e^0 = 1$$

and

$$a^{-n} = \frac{1}{a^n} = \frac{1}{e^{n\ln a}} = e^{-n\ln a} \quad \text{for} \quad n \in \mathbb{N}.$$

Thus equation (3) holds for all integers n. This allows us to denote the function $e^{x\ln a}$ by the symbol a^x, $a > 0$. We thus have

(4) $$a^x = e^{x\ln a}$$

for all $x \in \mathbb{R}$. Replacing in (2) x by $x\ln a$ and y by $y\ln a$ we get, in view of the last equality,

(5) $$a^x a^y = a^{x+y}.$$

From (1) and (4) we obtain

$$\left(e^{\ln a}\right)^x = e^{x\ln a}.$$

This equality holds for any real values $\ln a$ and x. In particular, replacing x by y and $\ln a$ by $x\ln a$ we get $\left(e^{x\ln a}\right)^y = e^{xy\ln a}$; that is,

(6) $$(a^x)^y = a^{xy}.$$

Equalities (5) and (6) hold for all x and y and represent fundamental properties of exponential functions.

If in (5) we substitute for y a natural number n and for x the number $-n$, then we have $a^{-n}a^{n} = 1$, whence

$$a^{-n} = \frac{1}{a^{n}},$$

(in accordance with the definition given in Section 4.14). This formula reduces the calculation of a^{-n} to multiplication and division, for a^{n} can be regarded as a product. Similarly, substituting in (6) a natural number n for y and the number $1/n$ for x we obtain

$$(a^{1/n})^{n} = a.$$

Hence we see that if a is a positive number and n is a natural number, then there always exists a positive number b such that $b^{n} = a$. That number is called the nth *principal root* of a and is often denoted by $\sqrt[n]{a}$. Thus

$$a^{1/n} = \sqrt[n]{a}.$$

Exercise 4.19

1.   Prove that the exponential function a^{x} is increasing if $a > 1$, and decreasing if $0 < a < 1$ (see Fig. 4.4).

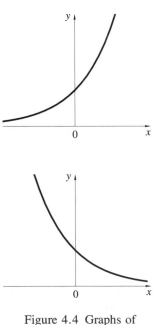

Figure 4.4 Graphs of $y = a^{x}$ for $a > 1$ and $0 < a < 1$.

4.20 THE NUMBER *e*

In the symbol a^x the letter a may denote an arbitrary positive number and x any arbitrary number at all. But when introducing the symbol e^x we emphasized the fact that the letter e has no separate meaning. Now we can say the e represents a number, namely $e = e^1$. The function e^x is thus a special case of the function a^x.

Since e is the value of e^x at $x = 1$, it is the sum of the series

(1) $$e = 1 + \frac{1}{1!} + \frac{1}{2!} + \frac{1}{3!} + \cdots .$$

The value of e can be found with arbitrary accuracy from (1). We have the inequalities

$$\sum_{k=0}^{n} \frac{1}{k!} < e < \sum_{k=0}^{n} \frac{1}{k!} + \frac{1}{n!n} .$$

(Recall that $0! = 1$.) The first inequality is obvious. To prove the second inequality let us remark that, for $k \geq n + 1$,

$$\frac{1}{k!} < \frac{1}{(n+1)!(n+1)^{k-n-1}}$$

and, consequently,

$$\sum_{k=n+1}^{\infty} \frac{1}{k!} < \frac{1}{(n+1)!} \sum_{k=n+1}^{\infty} \frac{1}{(n+1)^{k-n-1}} = \frac{1}{(n+1)!} \cdot \frac{1}{1 - (n+1)^{-1}} = \frac{1}{n!n} .$$

Taking $n = 13$ we have

(3) $$\sum_{k=0}^{13} \frac{1}{k!} = 1 + \frac{1}{1!} + \frac{1}{2!} + \frac{1}{3!} + \frac{1}{4!} + \frac{1}{5!} + \frac{1}{6!} + \frac{1}{7!} + \frac{1}{8!} + \frac{1}{9!} + \frac{1}{10!} + \frac{1}{11!} + \frac{1}{12!} + \frac{1}{13!} .$$

The fractions on the right side of (3) can easily be found one after the other with arbitrary accuracy. For instance, to calculate $1/7!$ it suffices to divide the already found value $1/6!$ by 7. In this way in few minutes we will have all the values:

$$
\begin{aligned}
1 &= 1 \\
1/1! &= 1 \\
1/2! &= 0.5 \\
1/3! &\approx 0.16666666666 \\
1/4! &\approx 0.04166666666 \\
1/5! &\approx 0.00833333333 \\
1/6! &\approx 0.00138888888 \\
1/7! &\approx 0.00019841269 \\
1/8! &\approx 0.00002480158 \\
1/9! &\approx 0.00000275573 \\
1/10! &\approx 0.00000027557 \\
1/11! &\approx 0.00000002505 \\
1/12! &\approx 0.00000000208 \\
1/13! &\approx 0.00000000016 \\
\hline
& 2.71828182839
\end{aligned}
$$

The numbers $1/3!,\ 1/4!,\ \ldots,\ 1/13!$ are calculated above with an error less than $1/10^{11}$. Thus their sum $S = 2.71828182839$ differs from (3) by at most $11/10^{11}$ (because we have added 11 approximate numbers). Consequently

$$S < e < S + 11/10^{11} + 1/(13!13) < S + 13/10^{11} < S + 1/10^{9},$$

and thus we may write

$$e \approx 2.718281828.$$

This number is easy to remember, because the sequence 1828 appears twice in it.

Exercise 4.20

1. Using the above method calculate $\sqrt[5]{e}$ and $1/\sqrt{e}$ with an error less than $1/10^{6}$.

4.21 INVERSE FUNCTIONS

As shown in Section 4.18, for each positive number x there exists exactly one number y such that

(1) $e^{y} = x.$

Thus this equality defines a function on $(0,\ \infty)$ which is called the *natural logarithm function* (or simply the *logarithm function*) and is denoted by

(2) $y = \ln x.$

The graph of the logarithm function (see Fig. 4.5) can be obtained by reflecting the graph of the function $y = e^{x}$ symmetrically with respect to the line $y = x$. The functions $y = e^{x}$ and $y = \ln x$ are inverse to each other. We are going to introduce the general concept of inverse functions.

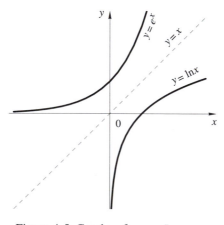

Figure 4.5 Graphs of $y = e^{x}$, $y = x$, and $y = \ln x$.

Let f be a function defined on a set F. Assume that for any two distinct points x_1 and x_2 of F the values $f(x_1)$ and $f(x_2)$ are different. Let G be the set of all values of f. In other words, G is the range of f. Then for every number y in G there is exactly one number x in F such that

$$(3) \qquad f(x) = y.$$

Thus the relation defines a function

$$(4) \qquad x = g(y)$$

whose domain is G and the range is F. The function g is called *inverse* to f. The inverse of f is often denoted by f^{-1}. This should not be confused with $1/f$.

If f is regarded as a set of ordered pairs, then the inverse function is obtained by interchanging the order of the elements in each pair. It is easily seen then that the concept of inverse function is symmetric: if g is inverse to f, then f is inverse to g. The graphs of inverse functions are symmetric with respect to the line $y = x$. The graph of f is the set of points $(x, f(x))$ on the plane. Consequently, the graph of the inverse function of f is the set of points $(f(x), x)$.

Substituting (3) in (4) we get

$$g(f(x)) = x \quad \text{for all } x \in F.$$

Similarly, substituting (4) in (3) we get

$$f(g(y)) = y \quad \text{for all } y \in G.$$

In other words, the composition of a function f and its inverse is the identity function on the domain of f. For instance,

$$e^{\ln x} = x \quad \text{for all } x > 0,$$

and

$$\ln e^x = x \quad \text{for all } x \in \mathbb{R}.$$

For some functions an inverse may not exist. If at two different points x_1 and x_2 the function has the same value y_0, then an inverse cannot be defined at y_0. A function that takes different values at different points is called *invertible*, or *one-to-one*.

Every increasing or decreasing function is invertible. Consequently, such functions have inverse functions.

4.21.1 Theorem *If a function f is increasing and continuous on an interval A, then the set B of all values of f (the range of f) is also an interval and the inverse of f is an increasing and continuous function on B. The theorem remains true if the word "increasing" is everywhere replaced by "decreasing."*

Proof. If y_1 and y_2 belong to B then f assumes every value between y_1 and y_2 which proves that B is an interval. Assume that $y_1 < y_2$. Then there exist two points x_1 and x_2 in A such that $f(x_1) = y_1$ and $f(x_2) = y_2$. Thus $f(x_1) < f(x_2)$. If f is increasing, we must have $x_1 < x_2$, which proves that the inverse function is increasing. If f is decreasing, then $x_1 > x_2$, which proves that the inverse function is decreasing.

It remains to prove that the inverse function is continuous. Assume that f is increasing and denote its inverse by g. Let y_0 be an interior point of B. Then the point $x_0 = g(y_0)$ lies in the interior of A. Given any $\varepsilon > 0$ we choose in A two numbers x_1 and x_2 such that

(5) $$x_0 - \varepsilon < x_1 < x_0 < x_2 < x_0 + \varepsilon.$$

If $y_1 = f(x_1)$ and $y_2 = f(x_2)$, then we have $y_1 < y_0 < y_2$. We can find a number $\delta > 0$ such that

$$y_1 < y_0 - \delta < y_0 + \delta < y_2.$$

If $|y - y_0| < \delta$, then $y_1 < y < y_2$ and $g(y_1) < g(y) < g(y_2)$, that is, $x_1 < g(y) < x_2$. Consequently, in view of (5), $x_0 - \varepsilon < g(y) < x_0 + \varepsilon$, which proves the continuity of g at y_0.

Exercises 4.21

1. Can you find an example of a function f for which the inverse f^{-1} equals $1/f$?
2. Prove that the composition of two invertible functions f and g is invertible and that its inverse is given by
$$(f \circ g)^{-1} = g^{-1} \circ f^{-1}.$$
3. True or false? If f and g are invertible, then $f + g$ is invertible.
4. True or false? If f and g are invertible, then fg is invertible.
5. Prove that if f is an invertible function and $f(x) \neq 0$ for all x in the domain of f, then the function $g(x) = 1/f(x)$ is invertible.

4.22 DERIVATIVE OF AN INVERSE FUNCTION

If f and g are inverse to each other, then $f(g(x)) = x$. If both f and g are differentiable, then by applying the theorem on the differentiation of composite functions we obtain $f'(g(x))g'(x) = 1$, and hence

$$g'(x) = \frac{1}{f'(g(x))}.$$

This formula shows how to find the derivative of an inverse function. It has been deduced under the assumption that both functions are differentiable. However, it suffices to assume only that one of them is.

4.22.1 Theorem (on differentiation of inverse functions) *If a function f is continuous and increasing (or decreasing) on an interval (a, b) and, moreover, has a derivative different from zero at a point $t \in (a, b)$, then the inverse function g has a derivative at the point $x = f(t)$ and this derivative is given by*

$$g'(x) = \frac{1}{f'(g(x))}.$$

Proof. Let us denote by t_0 a point at which the function f has the derivative different from 0 and assume that $x_0 = f(t_0)$. Define an auxiliary function

$$\rho(t) = \begin{cases} \dfrac{f(t) - f(t_0)}{t - t_0} - f'(t_0) & \text{if } t \neq t_0 \\ 0 & \text{if } t = t_0. \end{cases}$$

Then, since $t_0 = g(x_0)$,

$$x - x_0 = f(g(x)) - f(g(x_0)) = (g(x) - g(x_0))(f'(g(x_0)) + \rho(g(x))),$$

and hence for $x \neq x_0$,

(6) $$\frac{g(x) - g(x_0)}{x - x_0} = \frac{1}{f'(g(x_0)) + \rho(g(x))}.$$

If $x \to x_0$, then $\rho(g(x)) \to 0$, because g is continuous and equal to t_0 at x_0. Thus letting $x \to x_0$ in (6) yields $g'(x_0) = 1/f'(g(x_0))$, proving the theorem.

4.23 DIFFERENTIALS

—In some books the derivative is denoted by dy/dx. Is this the same derivative?

—Yes, it is. That notation goes back to the German mathematician Gottfried Wilhelm Leibniz (1646–1716). The advantage of this notation is that, for example, the formula in the Theorem 4.22.1 can be written as

$$\frac{dy}{dx} = \frac{1}{\dfrac{dx}{dy}}.$$

—I like this notation, because the formula is now easy to remember.

—That is true. But it still requires a proof.

—Why? The proof follows from the general rules of division.

—No, the rules of division have been stated for numbers and dx and dy are not numbers.

—So what are they?

—They actually have no separate meaning. An explanation of their use may be the symbol $\Delta y/\Delta x$, where $\Delta y = f(x) - f(x_0)$ and $\Delta x = x - x_0$. If x approaches x_0, then the differences become small, infinitely small and they are then called differentials.

—What is the definition of a differential?

—Leibniz probably imagined dx and dy as infinitely small numbers. Since the times of Leibniz the language of mathematics has changed. It became much more precise and logical. Expressions like "an infinitely small number" cannot be accepted as definitions. There exists a theory of infinitesimals, but this theory is rather difficult and sophisticated and, at least in calculus, introduces only complications.

In calculus textbooks a differential is defined as

$$dy = f'(x)dx,$$

where dx is an arbitrary number. So, for a fixed x, dy is regarded as a function of dx.

—Dividing both sides of the last inequality by dx we obtain

$$\frac{dy}{dx} = f'(x),$$

but I do not see any use of this manipulation.

—Neither do I.

—The chain rule in Section 4.16 can also be written as an algebraic identity

$$\frac{df}{dx} = \frac{df}{dy} \cdot \frac{dy}{dx}.$$

—Yes, but it also needs a proof. In any case it is worthwhile as a mnemonic way to remember the rule.

4.24 NATURAL LOGARITHM

The natural logarithm function $\ln x$ is inverse to e^x. It is defined on the interval $(0, \infty)$ and is continuous and increasing on this interval (Fig. 4.5). For $0 < x < 1$ it is negative and for $x > 1$ it is positive. Moreover $\ln 1 = 0$. The derivative of the natural logarithm function can be found by the formula given in Theorem 4.22.1. Since $(e^x)' = e^x$, we have

$$(\ln x)' = \frac{1}{e^{\ln x}} = \frac{1}{x}.$$

The function $\ln(x + 1)$ is defined on the interval $(-1, \infty)$. On the subinterval $(-1, 1)$ it can be expanded in a power series which can be found in the following way. First note that

(1)
$$\frac{1}{1 + x} = 1 - x + x^2 - x^3 + \cdots = \sum_{k=0}^{\infty} (-1)^k x^k,$$

where the series is convergent on the interval $(-1, 1)$. Since, for all $x \in (-1, 1)$,

$$(\ln(x + 1))' = \frac{1}{1 + x} = \sum_{k=0}^{\infty} (-1)^k x^k$$

and, by Theorem 4.13.2,

$$\left(\sum_{k=1}^{\infty} \frac{(-1)^{k+1}}{k} x^k \right)' = \sum_{k=0}^{\infty} (-1)^k x^k,$$

we must have

(2)
$$\ln(x + 1) = C + \sum_{k=1}^{\infty} \frac{(-1)^{k+1}}{k} x^k,$$

for some constant C and all $x \in (-1, 1)$. It remains to find the constant C. We can do that by substituting $x = 0$. This shows that $C = 0$. Consequently

$$(3) \quad \ln(x + 1) = \sum_{k=1}^{\infty} \frac{(-1)^{k+1}}{k} x^k = \frac{x}{1} - \frac{x^2}{2} + \frac{x^3}{3} - \frac{x^4}{4} + \cdots, \quad \text{for} \quad -1 < x < 1.$$

In Section 5.7 we shall prove that the equality holds also for $x = 1$. However, the considered series is divergent for $x = -1$ and all $|x| > 1$.

Exercises 4.24

1. Find the derivatives of the following functions:

   (a) $x^2 \ln(x^3)$;
   (b) $\ln(1 + x^2)$;

   (c) $\ln \dfrac{1 + x}{1 - x}$;
   (d) $e^x \ln x + \ln 3$;

   (e) $\ln(\ln x))$;
   (f) $\ln(1 + e^{-x})$.

2. Use formula (3) to calculate an approximate value of $\ln \frac{1}{2}$.

4.25 ARBITRARY LOGARITHMS

By differentiating the equality

$$a^x = e^{x \ln a},$$

(see Section 4.19) we get $(a^x)' = e^{x \ln a} \ln a = a^x \ln a$. We thus have

$$(a^x)' = a^x \ln a.$$

Hence it follows that the derivative is positive if $a > 1$ and negative if $0 < a < 1$. Thus the function a^x is increasing if $a > 1$ and decreasing if $0 < a < 1$. If $a = 1$, then $a^x = 1$.

The inverse function of a^x is called the *logarithm with base a* and is denoted by $\log_a x$. We always assume that a is positive and different from 1. The function $\log_a x$ is increasing if $a > 1$ and decreasing if $0 < a < 1$ (see Fig. 4.6). In both cases we have $\log_a a = 1$.

For all positive numbers x, y, and a ($a \neq 1$) the following equalities hold:

$$\log_a xy = \log_a x + \log_a y;$$

$$\log_a \frac{x}{y} = \log_a x - \log_a y;$$

$$\log_a x^p = p \log_a x \ (p \text{ is an arbitrary number});$$

$$\log_a x = \frac{\ln x}{\ln a};$$

$$(\log_a x)' = \frac{1}{x \ln a}.$$

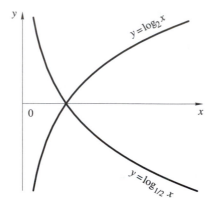

Figure 4.6 Graphs of $y = \log_a x$ for $a > 1$ and $0 < a < 1$.

The proofs run as follows:

$$\log_a xy = \log_a\left(a^{\log_a x} a^{\log_a y}\right) = \log_a\left(a^{\log_a x + \log_a y}\right) = \log_a x + \log_a y,$$

$$\log_a \frac{x}{y} = \log_a\left(\frac{a^{\log_a x}}{a^{\log_a y}}\right) = \log_a\left(a^{\log_a x - \log_a y}\right) = \log_a x - \log_a y,$$

$$\log_a x^p = \log_a\left(\left(a^{\log_a x}\right)^p\right) = \log_a\left(a^{p \log_a x}\right) = p \log_a x,$$

$$\log_a x = \frac{\log_a x \ln a}{\ln a} = \frac{\ln\left(a^{\log_a x}\right)}{\ln a} = \frac{\ln x}{\ln a},$$

$$(\log_a x)' = \left(\frac{\ln x}{\ln a}\right)' = \frac{1}{x \ln a}.$$

Since $\ln e = 1$, the equality $\log_a x = \ln x / \ln a$ implies that

$$\log_e x = \ln x.$$

Thus all the above formulas hold for the natural logarithm $\ln x$.

Exercises 4.25

1. Prove the formula $\log_a b \, \log_b a = 1$ for all positive a and b different from 1.
2. Find the derivatives of the following:
    (a) $a(x) = 2^x$.
    (b) $b(x) = \log_2(1 + x)$.
    (c) $c(x) = \ln \dfrac{1}{1 + 3^x}$,
    (d) $d(x) = 2^{(2^x)}$.

3. Prove the equality $(\log_x a)' = -\frac{(\log_x a)^2}{x \ln a}$, where a is positive and different from 1.

4.26 DERIVATIVES OF POWERS

By a *power* we mean the function

$$f(x) = x^a,$$

where the exponent a is an arbitrary number (Fig. 4.7). The case when a is an integer was already discussed in Sections 4.14 and 4.19. When a is not an integer, then we restrict the domain of the function to positive values of x.

By representing the power in the form

$$x^a = e^{a \ln x},$$

we easily find its derivative

$$(x^a)' = e^{a \ln x} \frac{a}{x} = x^a \frac{a}{x} = ax^{a-1}.$$

Hence the formula

$$(x^a)' = ax^{a-1}, \quad x > 0.$$

In particular we have

$$(x^{1/n})' = \frac{1}{n} x^{-(n-1)/n} \quad \text{and} \quad (x^{-1/n})' = -\frac{1}{n} x^{-(n+1)/n},$$

which can be written equivalently as

$$(\sqrt[n]{x})' = \frac{1}{n \sqrt[n]{x^{n-1}}} \quad \text{and} \quad \left(\frac{1}{\sqrt[n]{x}}\right)' = -\frac{1}{n \cdot \sqrt[n]{x^{n+1}}}.$$

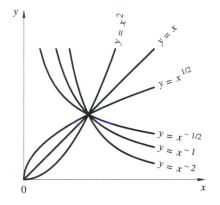

Figure 4.7 Examples of graphs of
$y = x^\alpha$ for different α.

In applications we use most often the square root, for which we have

$$(\sqrt{x})' = \frac{1}{2\sqrt{x}}.$$

It is worthwhile to keep this formula in mind. Equally important is the formula

$$\left(\ln(x + \sqrt{x^2 + a})\right)' = \frac{1}{\sqrt{x^2 + a}}, \quad \text{for} \quad x^2 + a > 0.$$

Indeed, we have

$$\left(\ln(x + \sqrt{x^2 + a})\right)' = \frac{\left(x + \sqrt{x^2 + a}\right)'}{x + \sqrt{x^2 + a}} = \frac{1 + \dfrac{(x^2 + a)'}{2\sqrt{x^2 + a}}}{x + \sqrt{x^2 + a}}$$

$$= \frac{x + \sqrt{x^2 + a}}{\sqrt{x^2 + a}\left(x + \sqrt{x^2 + a}\right)} = \frac{1}{\sqrt{x^2 + a}}.$$

It is important to remember that the power x^a is different from the exponential function a^x. However, the derivatives of both functions were found in the same way by representing them as $x^a = e^{a \ln x}$ and $a^x = e^{x \ln a}$.

The same method can be also used when the base and the exponents are arbitrary functions:

$$F(x) = g(x)^{f(x)}.$$

Namely, we can write

$$F = e^{f \ln g},$$

and hence

$$F' = e^{f \ln g}(f \ln g)' = g^f\left(\frac{f}{g}g' + f' \ln g\right).$$

This formula need not be remembered, because in practice it is more convenient to apply the described method in each case separately.

4.26.1 Examples

1. $(x^x)' = (e^{x \ln x})' = e^{x \ln x}(x \ln x)' = x^x(\ln x + 1), \quad x > 0.$
2. $((\ln x)^{x^2})' = (e^{x^2 \ln \ln x})' = e^{x^2 \ln \ln x}(x^2 \ln \ln x)' = (\ln x)^{x^2}(2x \ln \ln x + x/\ln x), \quad x > 0.$ ■

Exercises 4.26

1. Find the derivatives of the following functions:

   (a) $a(x) = \ln \dfrac{x}{1 + x}.$

   (b) $b(x) = \ln \dfrac{\sqrt{1 + x} - \sqrt{1 - x}}{\sqrt{1 + x} + \sqrt{1 - x}}.$

   (c) $c(x) = 3^x + x^3 + x^x + 3^3.$

   (d) $d(x) = x^{-2x}.$

(e) $e(x) = \dfrac{(8x^2 + 4x + 3)\sqrt{x^2 - 1}}{15x^2}.$ (f) $f(x) = \dfrac{2 + \sqrt{x}}{2 - \sqrt{x}}.$

(g) $g(x) = 2^{\sqrt{1+2x}}.$ (h) $h(x) = (\log_2 x)^{(e^x)}.$

2. In what intervals are the following functions increasing or decreasing?

(a) $a(x) = xe^{-x}.$

(b) $b(x) = (\ln x)^2 - \ln x^2.$

(c) $c(x) = x \cdot \sqrt{\dfrac{2 - x}{2 + x}}.$

4.27 NATURAL LOGARITHMS OF NATURAL NUMBERS

Formula (3) of Section 4.24,

$$\ln(1 + x) = \frac{x}{1} - \frac{x^2}{2} + \frac{x^3}{3} - \frac{x^4}{4} + \cdots = \sum_{n=1}^{\infty} \frac{(-1)^{n+1}}{n} x^n, \quad \text{for } -1 < x < 1,$$

can be used to calculate logarithms, but only for values that are close to 0. If we want to calculate the logarithm of natural numbers, we have to use a different formula, which can be obtained from the one above. First we replace x by $-x$:

$$\ln(1 - x) = -\frac{x}{1} - \frac{x^2}{2} - \frac{x^3}{3} - \frac{x^4}{4} + \cdots = -\sum_{n=1}^{\infty} \frac{x^n}{n}, \quad \text{for } -1 < x < 1.$$

Subtracting the last equality from the preceding one we get

(1) $\ln \dfrac{1 + x}{1 - x} = 2\left[\dfrac{x}{1} + \dfrac{x^3}{3} + \dfrac{x^5}{5} + \dfrac{x^7}{7} + \cdots\right] = 2\sum_{n=1}^{\infty} \dfrac{x^{2n-1}}{2n - 1}, \quad \text{for } -1 < x < 1.$

Now, for a given natural number p we select x such that

$$\frac{1 + x}{1 - x} = \frac{p + 1}{p};$$

thus $x = 1/(2p + 1)$. Substituting this value in (1) we obtain

$$\ln \frac{p + 1}{p} = 2\left[\frac{1}{1(2p + 1)} + \frac{1}{3(2p + 1)^3} + \frac{1}{5(2p + 1)^5} + \cdots\right]$$

$$= 2\sum_{n=1}^{\infty} \frac{1}{(2n - 1)(2p + 1)^{2n-1}}.$$

Adding $\ln p$ to both sides we finally obtain

(2) $\ln(p + 1) = \ln p + 2\left[\dfrac{1}{1(2p + 1)} + \dfrac{1}{3(2p + 1)^3} + \dfrac{1}{5(2p + 1)^5} + \cdots\right]$

$$= \ln p + 2\sum_{n=1}^{\infty} \frac{1}{(2n - 1)(2p + 1)^{2n-1}}.$$

Thus we can find the logarithm of $p + 1$ if we already know the logarithm of p. In this way we can successively calculate logarithms of natural numbers starting from the known equality $\ln 1 = 0$. For $p = 1$ we get from (2)

$$(3) \qquad \ln 2 = 2\left[\frac{1}{1\cdot 3} + \frac{1}{3\cdot 3^3} + \frac{1}{5\cdot 3^5} + \cdots\right] = 2\sum_{n=1}^{\infty}\frac{1}{(2n-1)3^{2n-1}}.$$

We have

$1/3$	≈ 0.33333333333	$1/(1\cdot 3)$	≈ 0.33333333333
$1/3^3$	≈ 0.03703703703	$1/(3\cdot 3^3)$	≈ 0.01234567901
$1/3^5$	≈ 0.00411522633	$1/(5\cdot 3^5)$	≈ 0.00082304526
$1/3^7$	≈ 0.00045724737	$1/(7\cdot 3^7)$	≈ 0.00006532105
$1/3^9$	≈ 0.00005080526	$1/(9\cdot 3^9)$	≈ 0.00000564502
$1/3^{11}$	≈ 0.00000564502	$1/(11\cdot 3^{11})$	≈ 0.00000051318
$1/3^{13}$	≈ 0.00000062722	$1/(13\cdot 3^{13})$	≈ 0.00000004824
$1/3^{15}$	≈ 0.00000006966	$1/(15\cdot 3^{15})$	≈ 0.00000000464
$1/3^{17}$	≈ 0.00000000774	$1/(17\cdot 3^{17})$	≈ 0.00000000045

The numbers of the first column are successively found by dividing the preceding number by 9. The numbers of the second column are found by dividing the numbers of the first column by $1, 3, 5, \ldots$, respectively. The sum of the right column is $S = 0.34657359018$. It is an approximate value of the sum of the 9 initial terms in (3). The possible error is less than $9/10^{11}$. Because all numbers are rounded down, we have

$$2S \le \ln 2 \le 2S + \frac{18}{10^{11}} + 2T,$$

where T is the sum of the neglected part of series (3). Since for $n \ge 9$

$$\frac{1}{(2n+1)\cdot 3^{2n+1}} \le \frac{1}{19\cdot 3^{19}}\frac{1}{9^{n-9}},$$

we have

$$T \le \frac{1}{19\cdot 3^{19}}\sum_{n=1}^{\infty}\frac{1}{9^{n-9}} = \frac{1}{19\cdot 3^{19}}\frac{1}{1-\frac{1}{9}} = \frac{1}{19\cdot 3^{17}\cdot 8} < \frac{51}{10^{11}}.$$

Hence

$$\ln 2 \approx 2S + \frac{18}{10^{11}} + \frac{102}{10^{11}} = 2S + \frac{120}{10^{11}} < 2S + \frac{1}{10^8}.$$

The above calculations show that 8 obtained digits are reliable, so that

$$\ln 2 \approx 0.69314718.$$

In practice the order of the calculations is slightly different. We first choose the degree of accuracy with which we want to find the logarithm. Then we calculate terms of the series with slightly more accuracy, usually by two or three more decimal places, because when summing approximate numbers the rate of error increases with every number summed. We evaluate consecutive terms of the series as long as we get nonzero numbers (within the accuracy chosen). An empiric method such as this allows us to establish, more or less, the number of terms we have to take into account. This method does not guarantee, however, that we actually obtain the degree of accuracy chosen so it is still necessary to estimate the error as we did in the above example.

To calculate $\ln 4$ we need not use the formula (2) because $\ln 4 = 2 \ln 2$. Thus we may write at once

$$\ln 4 \approx 1.38629436.$$

The number $\ln 5$ can be found from (2) when $p = 4$:

$$(4) \quad \ln 5 = \ln 4 + 2 \left[\frac{1}{1 \cdot 9} + \frac{1}{3 \cdot 9^3} + \frac{1}{5 \cdot 9^5} + \cdots \right] = \ln 4 + 2 \sum_{n=0}^{\infty} \frac{1}{(2n-1)9^{2n-1}}.$$

The sum of the first four terms

$$
\begin{aligned}
1/(1 \cdot 9) &\approx 0.1111111111 \\
1/(3 \cdot 9^3) &\approx 0.0004572473 \\
1/(5 \cdot 9^5) &\approx 0.0000033870 \\
1/(7 \cdot 9^7) &\approx 0.0000000295
\end{aligned}
$$

is $U = 0.1115717749$. We thus have

$$\frac{1}{1 \cdot 9} + \frac{1}{3 \cdot 9^3} + \frac{1}{5 \cdot 9^5} + \frac{1}{7 \cdot 9^7} < U + \frac{4}{10^{10}}.$$

The reader can verify that the sum of the neglected part of the series

$$\frac{1}{9 \cdot 9^9} + \frac{1}{11 \cdot 9^{11}} + \frac{1}{13 \cdot 9^{13}} + \cdots$$

is less than $4/10^{10}$. Hence

$$4S + 2U < \ln 5 < 2 \left(2S + \frac{12}{10^{10}} \right) + 2 \left(U + \frac{8}{10^{10}} \right)$$

$$= 4S + 2U + \frac{40}{10^{10}} < 4S + 2U + \frac{1}{10^8}.$$

This leads to the result

$$\ln 5 \approx 1.60943791.$$

The logarithm of 10 can be found from the equality $\ln 10 = \ln 2 + \ln 5$, whence we get

$$\ln 10 \approx 2.302585092.$$

4.28 COMMON LOGARITHMS

Logarithms change multiplication into addition according to the general formula

$$\log ab = \log a + \log b.$$

Since addition is much easier than multiplication, one can use logarithms in practical calculations. To this aim one needs extended tables which give logarithms of as many numbers as possible. But the range of such tables is always limited, so multiplication of large numbers is cumbersome. This difficulty can be evaded by introducing *common logarithms*. These are logarithms with base 10. They are also called decadic logarithms and can be calculated using the formula

$$\log_{10} x = \frac{\ln x}{\ln 10}.$$

Since

$$\frac{1}{\ln 10} \approx 0.4342944819,$$

the common logarithms can be obtained from natural logarithms just by multiplying them by the above constant factor.

—What is the advantage of introducing a new kind of logarithms that differ only by a constant factor?

—Tables of common logarithms need only contain numbers between 1 and 10, because each positive number can be expressed in the form

$$y = 10^n x,$$

where $1 < x < 10$. Clearly we have

$$\log_{10} y = n + \log_{10} x.$$

The number n is called the *characteristic* and the number $\log_{10} x$ the *mantissa* of the logarithm. The mantissa always lie between 0 and 1 so its digits begin just after the decimal point. The characteristic stands before the decimal point. For instance, $301 = 10^2 \cdot 3.01$, and hence

$$\log_{10} 301 \approx 2.47784$$

since $\log_{10} 3.01 \approx 0.47784$.

There have been great enthusiasts of common logarithms. The famous French mathematician Pierre Simon Laplace (1749–1827) said, "The invention of logarithms has shortened calculations extending over months to just a few weeks." Other mathematicians regarded the logarithmic tables as the very essence of mathematics. The first tables of mantissas were calculated by Henry Briggs (1561–1631).

Nowadays logarithmic tables have completely lost their importance, because they have been replaced by calculators which are much smaller, faster, and easier to use. However, sometimes even with a calculator in hand logarithms can prove useful. For example, try to calculate

$$\frac{999^{998}}{998^{999}}.$$

You cannot find 999^{998} and 998^{999} because these numbers are too large for a calculator. The answer can be easily obtained by using the following equality

$$\frac{999^{998}}{998^{999}} = e^{998\ln 999 - 999\ln 998}.$$

Exercise 4.28

1. Which number is bigger, 999^{998} or 998^{999}?

4.29 TRIGONOMETRIC FUNCTIONS AND GEOMETRY

Our aim is now to introduce trigonometric functions and establish connections between them.

—In most textbooks trigonometric functions are introduced in a geometric way. Those books appear in several editions and have been studied by generations of students.

—This is a bad situation. Analysis is based on a few axioms of numbers and each step is thoroughly justified. It is very inconsistent to use geometrical definitions there. This is a complete resignation of precision of exposition and a deviation from the analytic methods which make the core of calculus.

—Do you mean that the geometrical approach is not precise?

—It is not. However, it stimulates intuition, which is the first step before introducing precise methods.

—I see. If we wanted to use geometry for proofs in analysis, we should first state the axioms of geometry and establish their relationship to the axioms of numbers.

—Right. But even if we did so, the exposition would lose in its neatness and simplicity. It would also suggest that the axioms of geometry are necessary for introducing trigonometric functions.

—But in this book there is also a lot of geometry.

—Well, geometry is treated here as an illustration only and by no means as a proving instrument. A geometric point of view often helps in finding a logically correct, precise, analytical proof.

4.30 SINE AND COSINE

In physical applications the functions sine and cosine appear as solutions of the differential equation

(1) $$f''(x) = -f(x).$$

Assume that there exists a function f that satisfies this equation and that can be expanded in a power series:

$$f(x) = \sum_{k=0}^{\infty} a_k x^k.$$

Then we have in the interval of convergence

$$f'(x) = \sum_{k=0}^{\infty}(k + 1)a_{k+1}x^k,$$

$$f''(x) = \sum_{k=0}^{\infty}(k + 1)(k + 2)a_{k+2}x^k.$$

We see that (1) will be satisfied if we set

$$a_{k+2} = -\frac{a_k}{(k + 1)(k + 2)}$$

for all integers $k \geq 0$. In particular, we have

$$a_2 = -\frac{a_0}{2!}, \qquad a_4 = \frac{a_0}{4!}, \qquad a_6 = -\frac{a_0}{6!}, \ldots;$$

$$a_3 = -\frac{a_1}{3!}, \qquad a_5 = \frac{a_1}{5!}, \qquad a_7 = -\frac{a_1}{7!}, \ldots.$$

Let a_0 and a_1 be arbitrary fixed numbers and let A be the greater of $|a_0|$ and $|a_1|$. It is easy to prove by induction that

(2) $$|a_k| \leq \frac{A}{k!}$$

for all integers $k \geq 0$. As we know, the series

$$\left\{\sum_{k=0}^{n}\frac{A}{k!}|x|^k\right\}$$

converges for every $x \in \mathbb{R}$. Introducing the functions

(3) $$\cos x = 1 - \frac{x^2}{2!} + \frac{x^4}{4!} - \ldots = \sum_{k=0}^{\infty}(-1)^k\frac{x^{2k}}{(2k)!},$$

(4) $$\sin x = \frac{x}{1!} - \frac{x^3}{3!} + \frac{x^5}{5!} - \ldots = \sum_{k=0}^{\infty}(-1)^k\frac{x^{2k+1}}{(2k + 1)!},$$

we see that the defining series also converge for every $x \in \mathbb{R}$.

From (3) and (4) we can easily obtain the following equalities:

(5) $$(\cos x)' = -\sin x \quad \text{and} \quad (\sin x)' = \cos x;$$

(6) $$\cos 0 = 1 \quad \text{and} \quad \sin 0 = 0;$$

(7) $$\cos(-x) = \cos x \quad \text{and} \quad \sin(-x) = -\sin x.$$

The definitions of sine and cosine can be given in terms of equation (1).

4.30.1 Definition   The function f that satisfies the equation $f'' = -f$ and the conditions $f(0) = 0$, $f'(0) = 1$, is called the *sine function* and is denoted by $\sin x$.

Similarly, the function g that satisfies the equation $g'' = -g$ and the conditions $g(0) = 1$, $g'(0), = 0$ is called the *cosine function* and is denoted by $\cos x$.

This definition requires a proof of consistency, which would show that the functions sine and cosine not only exist but are unique. Their existence is shown above. Their uniqueness follows from general theorems in the theory of differential equations or can be proved directly, as we shall do in the next section.

Exercise 4.30

1. Prove (2).

4.31 UNIQUENESS OF THE DEFINITION OF SINE AND COSINE

In order to prove uniqueness of the functions f and g of Definition 4.30.1 we first show that if they satisfy

(1) $$f'' = -f \quad \text{and} \quad g'' = -g,$$

then

(2) $$f(0)g'(x + y) + f'(0)g(x + y) = f(x)g'(y) + f'(x)g(y)$$

for all $x, y \in \mathbb{R}$. In the proof we need not assume at all that the functions f and g can be expanded in power series.

We introduce an auxiliary function

(3) $$h(x) = f(x)g'(x_0 - x) + f'(x)g(x_0 - x),$$

where x_0 is an arbitrary fixed number. Then

$$h'(x) = -f(x)g''(x_0 - x) + f''(x)g(x_0 - x) = 0,$$

because of (1). This shows that h is a constant function. Hence $h(0) = h(x)$; that is, in view of (3),

$$f(0)g'(x_0) + f'(0)g(x_0) = f(x)g'(x_0 - x) + f'(x)g(x_0 - x).$$

The above equality holds for all $x, x_0 \in \mathbb{R}$. Thus, given any x and y we get the desired formula (2), by substituting $x_0 = x + y$.

Assume now that there are two solutions f_1 and f_2 of the equation

$$f'' = -f$$

satisfying the same initial conditions; that is, such that

$$f_1(0) = f_2(0) \quad \text{and} \quad f_1'(0) = f_2'(0).$$

Then the difference

$$h = f_1 - f_2$$

also satisfies the same equation. Letting $g(x) = \sin x$ we get from (2)

$$0 = h(x)\cos y + h'(x)\sin y,$$

since on the left side of (2) we have $h(0) = h'(0) = 0$. Putting $y = 0$ we get from the last equation $0 = h(x)$, or in other words, $f_1(x) = f_2(x)$. Thus the solution of Equation (1) is unique whenever the initial conditions are given. It hence follows that Definition 4.30.1 is consistent.

4.32 PROPERTIES OF SINE AND COSINE

—The functions sine and cosine used in geometry are different. They are functions of angles, whereas the arguments of the functions defined here are numbers.

—Right. But there is a simple connection between them. If we denote by $\text{Sin}\,\alpha$ and $\text{Cos}\,\alpha$ the functions of an angle α, then

$$\sin x = \text{Sin}\,\alpha \quad \text{and} \quad \cos x = \text{Cos}\,\alpha$$

whenever

$$\frac{x}{\pi} = \frac{\alpha}{180°}.$$

This relation is so simple that almost all the formulas for both kinds of functions are the same.

—I know that π is the ratio of the length of the circumference of a circle to its diameter. But can we give a precise analytical definition?

—The number $\pi/2$ is the least positive number x such that $\cos x = 0$.

—How do we know that such a number exists?

—We have $\cos 0 = 1$ and

$$\cos 2 = 1 - \frac{2^2}{2!} + \frac{2^4}{4!} - \left(\frac{2^6}{6!} - \frac{2^8}{8!}\right) - \cdots.$$

Since the sum of the three initial terms is $-\frac{1}{3}$ and all the remaining terms are negative, we have $\cos 2 < -\frac{1}{3}$. This implies, by Theorem 3.7.2, that $\cos x$ has a root in the interval $(0, 2)$. This root will be denoted by $\pi/2$.

—How do we know that there is only one root in $(0, 2)$?

—We have

$$\sin x = x\left(1 - \frac{x^2}{2 \cdot 3}\right) + \frac{x^5}{5!}\left(1 - \frac{x^2}{6 \cdot 7}\right) + \cdots,$$

which shows that $\sin x > 0$ on the interval $(0, 2)$. Since $-\sin x$ is the derivative of $\cos x$, it follows that $\cos x$ is decreasing in $(0, 2)$ and therefore it assumes the value 0 only once. So the number $\pi/2$ is defined uniquely.

—As we said, $\sin x$ and $\text{Sin}\,\alpha$ are different functions: for instance, $\sin \pi = 0$ and $\text{Sin}\,\pi \approx 0.0548$. Many authors consider both as the same function but defined on two different domains and they say "x radians" when the number x is the argument of the sine or the cosine. Such an approach is illogical. If the sine is a function of radians, then expressions like

$\sin x + x$ or $\sin(\sin x)$ make no sense. The concept of radians is completely unnecessary in mathematics.

Let us consider equality (2) of Section 4.31:

$$f(0)g'(x + y) + f'(0)g(x + y) = f(x)g'(y) + f'(x)g(y).$$

By letting $f(x) = \sin x$ and $g(x) = \cos x$ we get the following identity

(1) $$\cos(x + y) = \cos x \cos y - \sin x \sin y,$$

and by letting $f(x) = \sin x$ and $g(x) = \sin x$ we get

(2) $$\sin(x + y) = \sin x \cos y + \cos x \sin y.$$

Substituting $y = -x$ in (1) yields another trigonometric identity:

(3) $$1 = (\cos x)^2 + (\sin x)^2.$$

Since $\cos \frac{\pi}{2} = 0$ and $\sin \frac{\pi}{2} > 0$, the above implies that

$$\sin \frac{\pi}{2} = 1.$$

From (1) and (2) we obtain, successively,

$$\cos(x + \frac{\pi}{2}) = -\sin x, \qquad \sin(x + \frac{\pi}{2}) = \cos x,$$
$$\cos(x + \pi) = -\cos x, \qquad \sin(x + \pi) = -\sin x,$$
$$\cos(\pi - x) = -\cos x, \qquad \sin(\pi - x) = \sin x,$$
$$\cos(x + 2\pi) = \cos x, \qquad \sin(x + 2\pi) = \sin x.$$

The equalities in the last line say that the functions cosine and sine are periodic with period 2π. The approximate graphs of the cosine and sine functions look like those in Fig. 4.8.

Equalities (1), (2) and (3) are of crucial importance in trigonometry.

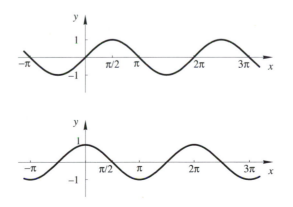

Figure 4.8 Graphs of $y = \sin x$ and $y = \cos x$.

Exercise 4.32

1. Prove the following:
    (a) $\cos x \cos y = \frac{1}{2} (\cos (x + y) + \cos (x - y))$.
    (b) $\cos x \sin y = \frac{1}{2} (\sin (x + y) - \sin (x - y))$.
    (c) $\sin x \sin y = \frac{1}{2} (\cos (x - y) - \cos (x + y))$.

4.33 TANGENT

The function *tangent* is defined by the formula

$$\tan x = \frac{\sin x}{\cos x};$$

its graph is represented in Fig. 4.9. This function is defined for all x for which $\cos x \neq 0$; that is, everywhere except $x = \frac{\pi}{2} + k\pi$, where k is an arbitrary integer. In particular

$$\tan 0 = 0 \quad \text{and} \quad \tan \frac{\pi}{4} = 1.$$

Moreover,

$$\tan x > 0 \quad \text{for } 0 < x < \frac{\pi}{2},$$

$$\tan x < 0 \quad \text{for } -\frac{\pi}{2} < x < 0,$$

and

$$\tan(-x) = -\tan x,$$
$$\tan(x + \pi) = \tan x,$$

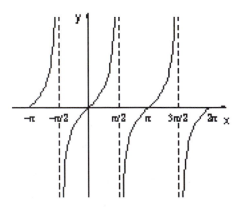

Figure 4.9 Graph of $f(x) = \tan(x)$.

because

$$\frac{\sin(x + \pi)}{\cos(x + \pi)} = \frac{-\sin x}{-\cos x}.$$

For the derivative of the tangent we have the formula

$$(\tan x)' = \frac{1}{(\cos x)^2},$$

which can be obtained by the theorem on differentiation of a quotient

$$\left(\frac{\sin x}{\cos x}\right)' = \frac{\cos x(\sin x)' - \sin x(\cos x)'}{(\cos x)^2} = \frac{(\cos x)^2 + (\sin x)^2}{(\cos x)^2} = \frac{1}{(\cos x)^2}.$$

The derivative is positive everywhere in the domain and the function is increasing on every interval not containing points $\frac{\pi}{2} + k\pi$.

The tangent function takes on all real values on $(-\frac{\pi}{2}, \frac{\pi}{2})$. In fact, if a is an arbitrary real number then there exist numbers x_1 and x_2 in that interval such that

$$\tan x_1 < a < \tan x_2,$$

because

$$\lim_{x \to -\frac{\pi}{2}+} \tan x = -\infty \quad \text{and} \quad \lim_{x \to \frac{\pi}{2}-} \tan x = \infty.$$

The assertion follows by Theorem 3.9.3
We also note that

(1)
$$\tan(x + y) = \frac{\tan x + \tan y}{1 - \tan x \tan y},$$

which holds for all x and y for which x, y, and $x + y$ are different from $\frac{\pi}{2} + k\pi$, where k is an integer.

Exercise 4.33

1. Prove (1).

4.34 A DIGRESSION ON GEOMETRY

In textbooks based on geometric definitions of the sine and cosine functions, the proof of the formula

(1)
$$\sin(\alpha + \beta) = \sin\alpha\cos\beta + \cos\alpha\sin\beta$$

is different from the one presented in the previous section. The standard proof is rather complicated. In this section we give an elegant and simple geometric proof of (1). We need only to know that the area of a triangle is equal to the product of the lengths of the sides adjacent to an angle γ times $\frac{1}{2}\sin\gamma$:

$$P = \frac{1}{2}ab\sin\gamma,$$

which can easily be seen in Fig. 4.10.

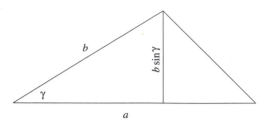

Figure 4.10 $P = \frac{1}{2}ab\sin\gamma$.

Assume now that $0 < \alpha < \frac{\pi}{2}$ and $0 < \beta < \frac{\pi}{2}$. Construct a triangle ABC (Fig. 4.11) such that the angle ACD is α, the angle BCD is β, AB is perpendicular to CD, and the length of CD is 1. Then by calculating the area of the triangle ABC in two different ways we obtain

$$\frac{1}{2}\frac{1}{\cos\alpha}\frac{1}{\cos\beta}\sin(\alpha+\beta) = \frac{1}{2}(\tan\alpha + \tan\beta).$$

Multiplying both sides by $2\cos\alpha\cos\beta$ we get equality (1).

4.35 ARCTANGENT

When introducing the function inverse to the tangent, we have to restrict the tangent's domain to an interval on which tangent is a one-to-one function. To this aim we choose the interval $(-\frac{\pi}{2}, \frac{\pi}{2})$. The function inverse to the tangent restricted to this interval is called the *arctangent* and is denoted arctan. The arctangent function is defined for all real numbers and its range is the interval $(-\frac{\pi}{2}, \frac{\pi}{2})$.

We have to remember that the function arctangent is not the inverse of the tangent, but the inverse of its restriction to the interval $(-\frac{\pi}{2}, \frac{\pi}{2})$ (see Fig. 4.12). The choice of that interval is a question of convention; instead we could take any interval on which the tangent is increasing.

From the properties of the tangent we easily obtain the following properties of the arctangent:

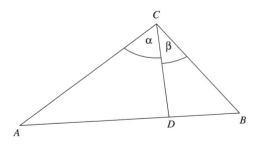

Figure 4.11 $\frac{1}{2}\frac{1}{\cos\alpha}\frac{1}{\cos\beta}\sin(\alpha+\beta) = \frac{1}{2}(\tan\alpha + \tan\beta)$.

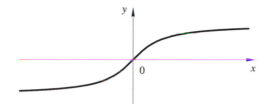

Figure 4.12 Graph of $y = \arctan x$.

$$\arctan 0 = 0, \qquad \arctan 1 = \frac{1}{4}\pi;$$

$$0 < \arctan x < \frac{\pi}{2} \quad \text{for } x > 0;$$

(1)
$$-\frac{\pi}{2} < \arctan x < 0 \quad \text{for } x < 0;$$

$$\arctan(-x) = -\arctan x \quad \text{for all } x.$$

According to the definition of the arctangent the equality $\arctan(\tan x) = x$ holds only for $-\frac{\pi}{2} < x < \frac{\pi}{2}$. On the other hand, the formula $\tan(\arctan x) = x$ holds for all x.

Of importance is the formula

(2)
$$\arctan x + \arctan y = \arctan \frac{x+y}{1-xy} \quad \text{for } 1 - xy > 0.$$

It can be derived from (1) of Section 4.33 as follows. First note that

(3)
$$\tan\!\left(\arctan x + \arctan y\right) = \frac{\tan(\arctan x) + \tan(\arctan y)}{1 - \tan(\arctan x)\tan(\arctan y)} = \frac{x+y}{1-xy}$$

for all x, y. Hence

(4)
$$\arctan\left[\tan\!\left(\arctan x + \arctan y\right)\right] = \arctan \frac{x+y}{1-xy}.$$

The left hand side can be replaced by $\arctan x + \arctan y$, provided that

(5)
$$-\frac{\pi}{2} < \arctan x + \arctan y < \frac{\pi}{2}.$$

We shall show that these inequalities hold whenever $1 - xy > 0$; that is, $xy < 1$. Indeed, since $0 < \arctan x < \frac{\pi}{2}$ for $x > 0$ and $-\frac{\pi}{2} < \arctan x < 0$ for $x < 0$, inequalities (5) hold if $xy \le 0$. If $xy > 0$, we consider two cases:

Case 1. $x > 0$ and $y > 0$. Then $0 < \arctan x + \arctan y < \pi$. Moreover, since $xy < 1$ we have $(x+y)/(1-xy) > 0$, and by (3), $\tan(\arctan x + \arctan y) > 0$. Therefore

$$0 < \arctan x + \arctan y < \frac{\pi}{2},$$

because in the interval $(\frac{\pi}{2}, \pi)$ the function tangent is negative.

Case 2. $x < 0$ and $y < 0$. Then $-\pi < \arctan x + \arctan y < 0$. Moreover, since $xy < 1$ we have $(x + y)/(1 - xy) < 0$, and by (3), $\tan(\arctan x + \arctan y) < 0$. Therefore

$$-\frac{\pi}{2} < \arctan x + \arctan y < 0,$$

because on the interval $(-\pi, -\frac{\pi}{2})$ the function tangent is positive.

We thus have proved that $1 - xy > 0$ implies (5). Consequently, formula (4) reduces to (2).

Exercises 4.35

1. Prove properties (1).
2. Prove that $\arctan x + \arctan(1/x) = \frac{\pi}{2}$ for all $x > 0$.

4.36 THE DERIVATIVE OF THE ARCTANGENT

For all $x \in \mathbb{R}$ we have

(1) $$(\arctan x)' = \frac{1}{1 + x^2}.$$

Indeed, using the general formula for the derivative of an inverse function we get

$$(\arctan x)' = (\cos(\arctan x))^2.$$

It thus suffices to prove that

(2) $$(\cos(\arctan x))^2 = \frac{1}{1 + x^2}.$$

Denoting $\arctan x$ by y we have $x = \tan y$ and

$$1 + x^2 = 1 + (\tan y)^2 = 1 + \left(\frac{\sin y}{\cos y}\right)^2 = \frac{(\cos y)^2 + (\sin y)^2}{(\cos y)^2} = \frac{1}{(\cos y)^2} = \frac{1}{(\cos(\arctan x))^2}.$$

This implies (2) and subsequently (1).

We are going to prove now the following equality

(3) $$\arctan x = \frac{x}{1} - \frac{x^3}{3} + \frac{x^5}{5} - \frac{x^7}{7} + \cdots = \sum_{k=0}^{\infty} (-1)^k \frac{x^{2k+1}}{2k + 1} \quad \text{for } |x| < 1.$$

The series is obviously convergent on $(-1, 1)$ because

$$\left| (-1)^k \frac{x^{2k+1}}{2k + 1} \right| \leq x^{2k}$$

and the series $\{\sum_{k=0}^{n} x^{2k}\}$ converges on $(-1, 1)$. Moreover,

$$\left(\sum_{k=0}^{\infty}(-1)^k \frac{x^{2k+1}}{2k+1}\right)' = \sum_{k=0}^{\infty}(-1)^k x^{2k} = 1 - x^2 + x^4 - x^6 + \cdots = \frac{1}{1+x^2}$$

for $|x| < 1$. Thus, in view of (1), for all $x \in (-1, 1)$ we must have

$$\arctan x = C + \frac{x}{1} - \frac{x^3}{3} + \frac{x^5}{5} - \frac{x^7}{7} + \cdots = C + \sum_{k=0}^{\infty}(-1)^k \frac{x^{2k+1}}{2k+1}$$

where C is a constant. This number can be found by substituting $x = 0$, which yields $C = 0$. This proves (3).

We shall prove in Section 5.7 that (3) holds also for $x = 1$.

The series $\{\sum_{k=0}^{n}(-1)^k x^{2k+1}/(2k+1)\}$ is an example of an alternating series, that is, its terms are alternatively positive and negative. When using such an expansion for practical calculations the error can be conveniently estimated by the following theorem, due to Leibniz. In calculus textbooks this theorem is usually called the *Alternating Series Test*.

4.36.1 Theorem (Leibniz's criterion) *If a sequence $a_0, a_1, a_2, \ldots$ is monotone decreasing to 0, then the series*

$$(4) \qquad \left\{\sum_{k=0}^{n}(-1)^k a_k\right\}$$

converges. Moreover, for each $n \in \mathbb{N}$,

$$(5) \qquad s_{2n-1} < \sum_{k=0}^{\infty}(-1)^k a_k < s_{2n}, \quad \text{where } s_m = \sum_{k=0}^{m}(-1)^k a_k.$$

Proof. Since the sequence $a_0, a_1, a_2, \ldots$ is decreasing to 0, we have

$$a_{2n} - a_{2n+1} > 0, \qquad a_{2n+2} > 0, \quad \text{and} \quad a_{2n+1} - a_{2n+2} > 0.$$

Hence

$$\begin{aligned} s_{2n-1} &< s_{2n-1} + (a_{2n} - a_{2n+1}) \\ &= s_{2n+1} < s_{2n+1} + a_{2n+2} \\ &= s_{2n+2} = s_{2n} - (a_{2n+1} - a_{2n+2}) < s_{2n} \end{aligned}$$

and consequently

$$s_{2n-1} < s_{2n+1} < s_{2n+2} < s_{2n}.$$

This shows that the sequences $\{s_{2n}\}$ and $\{s_{2n-1}\}$ are both monotone and bounded, and thus convergent. Now $s_{2n} - s_{2n-1} = a_{2n}$ implies that

$$\lim_{n\to\infty}(s_{2n} - s_{2n-1}) = \lim_{n\to\infty} a_{2n} = 0$$

or

$$\lim_{n\to\infty} s_{2n} = \lim_{n\to\infty} s_{2n-1}.$$

This proves convergence of the sequence $\{s_n\}$. Let $s = \lim_{n \to \infty} s_n$. Because the sequence $\{s_{2n}\}$ is decreasing, the sequence $\{s_{2n-1}\}$ is increasing, and both sequences converge to s, we have

$$s_{2n-1} < s < s_{2n},$$

proving (5).

Exercises 4.36

1. Find the derivatives of the following functions:

    (a) $a(x) = (\tan \sqrt{x})^4$.

    (b) $b(x) = \ln\left[\tan\left(\frac{1}{4}\pi + \frac{1}{2}x\right)\right]$.

    (c) $c(x) = (\tan x)^{\sin x}$.

    (d) $d(x) = x^4 \arctan x$.

    (e) $e(x) = \arctan \sqrt{1 - x^2}$.

2. Are the following series convergent?

    (a) $\left\{ \sum_{k=1}^{n} \frac{(-1)^{k+1}}{\ln(k + 1)} \right\}$;

    (b) $\left\{ \sum_{k=1}^{n} \frac{(-1)^{k+1}}{\arctan k} \right\}$;

    (c) $\left\{ \sum_{k=1}^{n} \frac{\cos k\pi}{k} \right\}$.

3. How many terms of the given series should one use to compute the sum with three-decimal-place accuracy?

    (a) $\left\{ \sum_{k=1}^{n} \frac{(-1)^{k+1}}{k} \right\}$;

    (b) $\left\{ \sum_{k=1}^{n} \frac{(-1)^{k+1}}{\ln(k + 1)} \right\}$;

    (c) $\left\{ \sum_{k=1}^{n} \frac{(-1)^{k+1}}{k!} \right\}$.

4.37 ESTIMATION OF π

Expansion (3) in Section 4.36 and formula (2) in Section 4.35 permit us to evaluate π with arbitrary accuracy. Indeed, we have

$$2 \arctan \frac{1}{5} = \arctan \frac{\frac{1}{5} + \frac{1}{5}}{1 - \frac{1}{25}} = \arctan \frac{5}{12},$$

$$4 \arctan \frac{1}{5} = \arctan \frac{\frac{5}{12} + \frac{5}{12}}{1 - \frac{25}{144}} = \arctan\left(1 + \frac{1}{119}\right),$$

$$4 \arctan \frac{1}{5} - \arctan 1 = \arctan \frac{\left(1 + \frac{1}{119}\right) - 1}{1 + \left(1 + \frac{1}{119}\right)} = \arctan \frac{1}{239}.$$

Since $\arctan 1 = \frac{1}{4}\pi$, the above equality implies that

$$\frac{1}{4}\pi = 4\arctan\frac{1}{5} - \arctan\frac{1}{239}.$$

Hence

(1) $\quad \pi = 16\left(\frac{1}{5} - \frac{1}{3\cdot 5^3} + \frac{1}{5\cdot 5^5} - \cdots\right) - 4\left(\frac{1}{239} - \frac{1}{3\cdot 239^3} + \cdots\right)$

$$= 16\sum_{n=0}^{\infty}\frac{(-1)^n}{(2n+1)5^{2n+1}} - 4\sum_{n=0}^{\infty}\frac{(-1)^n}{(2n+1)239^{2n+1}}.$$

In order to find an approximate value of π let us take six terms of the first series and only two terms of the second series. Then, according to Theorem 4.36.1, the error will be less than

(2) $\qquad\qquad 16\frac{1}{13\cdot 5^{13}} + 4\frac{1}{5\cdot 239^5} \approx 10^{-9}.$

The calculations run like this:

$$16\left(\frac{1}{5} - \frac{1}{3\cdot 5^3} + \frac{1}{5\cdot 5^5} - \frac{1}{7\cdot 5^7} + \frac{1}{9\cdot 5^9} - \frac{1}{11\cdot 5^{11}}\right) \approx 3.15832895662,$$

$$4\left(\frac{1}{239} - \frac{1}{3\cdot 239^3}\right) \approx 0.01673630401,$$

$$3.15832895662 - 0.01673630401 = 3.14159265261.$$

Since, by (2), eight digits are reliable, we can write

$$\pi \approx 3.14159265.$$

Formula (1) was given by John Machin (1685–1751), who found π with 100-place accuracy when he was 21 years old.

4.38 NEWTON'S BINOMIAL

The function $(1 + x)^\alpha$, where α is an arbitrary real number, is called *Newton's binomial*. It is defined for all $x > -1$. Assume that this function expands in a power series:

(1) $\qquad\qquad (1 + x)^\alpha = c_0 + c_1 x + c_2 x^2 + \cdots = \sum_{k=0}^{\infty} c_k x^k.$

In order to find the coefficients c_k we differentiate the above equality to obtain

$$\alpha(1 + x)^{\alpha-1} = c_1 + 2c_2 x + 3c_3 x^2 + \cdots = \sum_{k=1}^{\infty} k c_k x^{k-1}.$$

Hence

$$(1 + x)(c_1 + 2c_2x + 3c_3x^2 + \cdots) = \alpha(c_0 + c_1x + c_2x^2 + \cdots);$$

that is,

$$c_1 + 2c_2x + 3c_3x^2 + \cdots + c_1x + 2c_2x^2 + 3c_3x^3 + \cdots$$
$$= \alpha c_0 + \alpha c_1x + \alpha c_2x^2 + \alpha c_3x^3 + \cdots$$

We know that $c_0 = 1$. Comparing the coefficients in the above equality we find

$$c_1 = \frac{\alpha}{1}, \qquad c_2 = \frac{\alpha(\alpha - 1)}{1 \cdot 2}, \qquad c_3 = \frac{\alpha(\alpha - 1)(\alpha - 2)}{1 \cdot 2 \cdot 3},$$

and in general

$$(2) \qquad c_k = \frac{\alpha(\alpha - 1)\ldots(\alpha - k + 1)}{k!}.$$

To find the interval of convergence we shall use the following general theorem.

4.38.1 Theorem *If the limit*

$$\lim_{k \to \infty} \left| \frac{c_k}{c_{k+1}} \right| = r$$

exists, then r is the radius of convergence of the series

$$(3) \qquad \left\{ \sum_{k=0}^{n} c_k x^k \right\}.$$

Proof. Let x be an arbitrary fixed number such that $|x| < r$. For $|x| < q < r$ there exists a natural number p such that $|c_k/c_{k+1}| > q$ for all $k \geq p$; that is,

$$|c_{k+1}| < \frac{|c_k|}{q} \quad \text{for} \quad k = p, \ p + 1, \ p + 2, \ldots.$$

Hence

$$|c_{p+1}| < \frac{|c_p|}{q}, \qquad |c_{p+2}| < \frac{|c_{p+1}|}{q} < \frac{|c_p|}{q^2},$$

and in general, by induction,

$$|c_{p+k}| < \frac{|c_p|}{q^k} \quad \text{for } k = 1, 2, 3, \ldots.$$

Thus

$$|c_{p+k} x^{p+k}| \leq |c_p x^p| \left| \frac{x}{q} \right|^k \quad \text{for } k = 1, 2, 3, \ldots,$$

which implies the convergence of series (3) by comparison with the geometric series

$$\left\{ \sum_{k=0}^{n} |c_p x^p| \left| \frac{x}{q} \right|^k \right\}.$$

Now, let $|x| > r$. Then there exists a number q such that $r < q < |x|$. By an argument similar to the above we can show that

(4) $$\left| c_{p+k} x^{p+k} \right| \geq \left| c_p x^p \right| \left| \frac{x}{q} \right|^k \quad \text{for } k = 1, 2, 3, \ldots.$$

If series (3) were convergent at x, then also the geometric series

$$\left\{ \sum_{k=0}^{n} \left| c_p x^p \right| \left| \frac{x}{q} \right|^k \right\}$$

would be convergent, which is not true. Thus the radius of convergence of series (3) is r.

Theorem 4.38.1 will be now applied to our series (1). In this case we have

$$\lim_{k \to \infty} \left| \frac{c_k}{c_{k+1}} \right| = \lim_{k \to \infty} \left| \frac{k+1}{\alpha - k} \right| = 1.$$

Thus the radius of convergence is 1 and the series converges for every $x \in (-1, 1)$.
In particular, for $\alpha = -\frac{1}{2}$ we have

$$\frac{1}{\sqrt{1+x}} = 1 - \frac{1}{2}x + \frac{1 \cdot 3}{2 \cdot 4}x^2 - \frac{1 \cdot 3 \cdot 5}{2 \cdot 4 \cdot 6}x^3 + \cdots.$$

This formula can serve for the approximate evaluation of square roots. For instance, we may write

$$\sqrt{2} = \frac{7}{5}\left(1 - \frac{1}{50}\right)^{-\frac{1}{2}} = \frac{7}{5}\left(1 + \frac{1}{2} \cdot \frac{1}{50} + \frac{1 \cdot 3}{2 \cdot 4} \cdot \frac{1}{50^2} + \frac{1 \cdot 3 \cdot 5}{2 \cdot 4 \cdot 6} \cdot \frac{1}{50^3} + \cdots\right).$$

Faster convergence would be obtained by using

$$\sqrt{2} = \frac{141}{100}\left(1 - \frac{119}{20000}\right)^{-\frac{1}{2}}$$

or any other equality that we would obtain by taking a rough approximation α of $\sqrt{2}$ (1.41 in the above example) and writing

$$\sqrt{2} = \alpha\left(1 + \frac{\alpha^2 - 2}{2}\right)^{-\frac{1}{2}}.$$

Similarly, using the fact that $\sqrt{3} \approx 1.732$ we are lead to the equality

$$\sqrt{3} = 1.732\left(1 - \frac{176}{3000000}\right)^{-\frac{1}{2}},$$

which allows us to calculate $\sqrt{3}$ to 50 decimal places without much difficulty. Such accuracy is not available on ordinary calculators.

The same method can be also applied to roots of higher order. For example, we can use the equalities

$$\sqrt[3]{3} = \frac{10}{7}\left(1 + \frac{29}{1000}\right)^{\frac{1}{3}} \quad \text{or} \quad \sqrt[5]{1000} = 4\left(1 + \frac{24}{1000}\right)^{-\frac{1}{5}}.$$

Using the notation

$$\frac{\alpha(\alpha - 1)\cdots(\alpha - k + 1)}{1 \cdot 2 \cdots k} = \binom{\alpha}{k}$$

the *binomial formula* can be written in the form

$$(1 + x)^\alpha = \sum_{k=0}^{\infty} \binom{\alpha}{k} x^k \quad \text{for } |x| < 1,$$

where we let $\binom{\alpha}{0} = 1$.

If $\alpha = m$ is a natural number, then $\binom{m}{k} = 0$ for $k > m$ and consequently the last formula reduces to the polynomial

$$(1 + x)^m = \sum_{k=0}^{m} \binom{m}{k} x^k$$

and is valid for all x, because if two polynomials are equal for $|x| < 1$, they are equal for every x.

If a and b $(a \neq 0)$ are arbitrary real numbers, we may write

$$(a + b)^m = a^m\left(1 + \frac{b}{a}\right)^m = a^m \sum_{k=0}^{m} \binom{m}{k}\left(\frac{b}{a}\right)^k$$

and hence

$$(a + b)^m = \sum_{k=0}^{m} \binom{m}{k} a^{m-k} b^k,$$

or

(5) $$(a + b)^m = \binom{m}{0} a^m + \binom{m}{1} a^{m-1}b + \binom{m}{2} a^{m-2}b^2 + \cdots + \binom{m}{m} b^m.$$

Since the formula holds also for $a = 0$, it holds for any real numbers a and b.

For initial values of m we get

$$(a + b)^1 = a + b$$
$$(a + b)^2 = a^2 + 2ab + b^2$$
$$(a + b)^3 = a^3 + 3a^2b + 3ab^2 + b^3$$
$$(a + b)^4 = a^4 + 4a^3b + 6a^2b^2 + 4ab^3 + b^4$$

The coefficients can be arranged into the so-called *Pascal's triangle* (Blaise Pascal, 1623–1662):

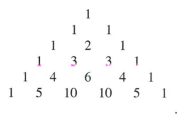

in which every entry is the sum of the two numbers standing just above.

From (5) we can easily get various interesting properties of the symbol $\binom{m}{k}$. Taking, for instance, $a = b = 1$, we have

$$2^m = \binom{m}{0} + \binom{m}{1} + \cdots + \binom{m}{m},$$

which would be difficult to deduce directly from the definition of the symbol $\binom{m}{k}$.

Exercises 4.38

1. Calculate $\sqrt{2}$, $\sqrt{3}$, and $\sqrt[3]{3}$ up to 6 decimal places and check the results by a calculator.

2. Prove the following:

   (a) $\binom{m}{0} - \binom{m}{1} + \binom{m}{2} - \ldots + (-1)^m \binom{m}{m} = 0.$

   (b) $\binom{m}{0} + \binom{m}{2} + \binom{m}{4} + \cdots + \binom{m}{m} = 2^{m-1}$ for even m.

3. Find the radii of convergence:

   (a) $\left\{ \sum_{k=1}^{n} \dfrac{k^2 x^k}{2^k} \right\};$ (b) $\left\{ \sum_{k=1}^{n} \dfrac{(-1)^k}{\sqrt{k}} x^k \right\};$

   (c) $\left\{ \sum_{k=1}^{n} (-3)^k x^{2k} \right\}.$

4.39 FUNCTIONS ARCCOSINE AND ARCSINE

The functions cosine and sine are invertible on any interval on which they are monotone. For the cosine function we choose the interval $[0, \pi]$ and for the sine function the interval $[-\frac{\pi}{2}, \frac{\pi}{2}]$. The inverse functions are called *arccosine* and *arcsine* and denoted by

$$\arccos x \quad \text{and} \quad \arcsin x,$$

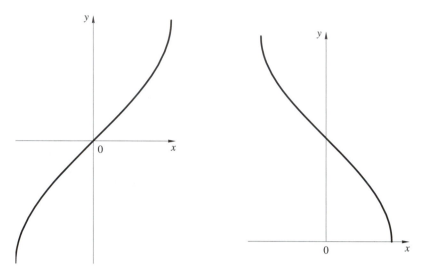

Figure 4.13 Graph of $y = \arcsin x$ and $y = \arccos x$.

respectively (see Fig. 4.13). Both are defined on the interval $[-1, 1]$ and satisfy the inequalities

$$0 \leq \arccos x \leq \pi \quad \text{and} \quad -\frac{\pi}{2} \leq \arcsin x \leq \frac{\pi}{2}.$$

Since

$$\sin(\arcsin x) = x = \cos(\arccos x) = \sin(\frac{\pi}{2} - \arccos x),$$

it follows $\arcsin x = \pi/2 - \arccos x$; that is,

(1) $$\arccos x + \arcsin x = \frac{\pi}{2} \quad \text{for} \quad -1 \leq x \leq 1.$$

We shall also prove the formulas

(2) $$(\arccos x)' = -\frac{1}{\sqrt{1 - x^2}} \quad \text{and} \quad (\arcsin x)' = \frac{1}{\sqrt{1 - x^2}}.$$

In view of (1) it suffices to prove one of these formulas. By the general theorem on the differentiation of inverse functions, we get

(3) $$(\arcsin x)' = \frac{1}{\cos(\arcsin x)}.$$

Denoting $\arcsin x$ by y we have $x = \sin y$ and hence

$$\sqrt{1 - x^2} = \sqrt{1 - (\sin y)^2} = \sqrt{(\cos y)^2} = \cos y,$$

because $\cos y \geq 0$ for $-\frac{\pi}{2} \leq y \leq \frac{\pi}{2}$. Thus

$$\sqrt{1 - x^2} = \cos(\arcsin x),$$

proving (2).

Now we are going to expand the arcsine function into a power series. From the binomial formula we have

$$(4) \qquad \frac{1}{\sqrt{1-x^2}} = 1 + \frac{1}{2}x^2 + \frac{1\cdot 3}{2\cdot 4}x^4 + \frac{1\cdot 3\cdot 5}{2\cdot 4\cdot 6}x^6 + \cdots$$

for $|x| < 1$. Moreover, for all $|x| < 1$,

$$\left(\frac{x}{1} + \frac{1}{2}\cdot\frac{x^3}{3} + \frac{1\cdot 3}{2\cdot 4}\cdot\frac{x^5}{5} + \frac{1\cdot 3\cdot 5}{2\cdot 4\cdot 6}\cdot\frac{x^7}{7} + \cdots\right)' = 1 + \frac{1}{2}x^2 + \frac{1\cdot 3}{2\cdot 4}x^4 + \frac{1\cdot 3\cdot 5}{2\cdot 4\cdot 6}x^6$$

$$+ \cdots = \frac{1}{\sqrt{1-x^2}}$$

(note that the differentiated series converges for $|x| < 1$). This implies that

$$\operatorname{arcsin} x = C + \frac{x}{1} + \frac{1}{2}\cdot\frac{x^3}{3} + \frac{1\cdot 3}{2\cdot 4}\cdot\frac{x^5}{5} + \frac{1\cdot 3\cdot 5}{2\cdot 4\cdot 6}\cdot\frac{x^7}{7} + \cdots \quad \text{for } |x| < 1.$$

The constant C can be easily established to be 0, because for $x = 0$ both functions vanish. Thus we have

$$\operatorname{arcsin} x = \frac{x}{1} + \frac{1}{2}\cdot\frac{x^3}{3} + \frac{1\cdot 3}{2\cdot 4}\cdot\frac{x^5}{5} + \frac{1\cdot 3\cdot 5}{2\cdot 4\cdot 6}\cdot\frac{x^7}{7} + \cdots \quad \text{for } |x| < 1.$$

Exercises 4.39

1. Find the power series for $\operatorname{arccos} x$.

2. Expand into a power series the function $\ln\!\left(x + \sqrt{1 + x^2}\right)$.

3. Find the derivatives of the following functions:

    (a) $a(x) = \operatorname{arcsin}\sqrt{x^3}$. \qquad (b) $b(x) = \operatorname{arccos}\sqrt{1-x}$.

    (c) $c(x) = x^3 \operatorname{arcsin}\frac{1}{x}$.

    (d) $d(x) = \dfrac{1}{x}\operatorname{arcsin}\sqrt{1-x^2} + \dfrac{1}{2}\ln\dfrac{1 - \sqrt{1-x^2}}{1 + \sqrt{1-x^2}}$.

4.40 COMPENDIUM OF FORMULAS

The skill of using differential calculus requires knowledge of derivatives of elementary functions:

$(c)' = 0$, c a constant. \qquad $(\sin x)' = \cos x$.

$(x^\alpha)' = \alpha x^{\alpha - 1}$. \qquad $(\cos x)' = -\sin x$.

$(e^x)' = e^x$. \qquad $(\tan x)' = \dfrac{1}{(\cos x)^2}$.

$(\alpha^x)' = \alpha^x \ln\alpha$. \qquad $(\operatorname{arcsin} x)' = \dfrac{1}{\sqrt{1-x^2}}$.

$$(\ln x)' = \frac{1}{x}. \qquad\qquad (\arccos x)' = -\frac{1}{\sqrt{1-x^2}}.$$

$$\left(\ln\left|x + \sqrt{\alpha + x^2}\right|\right)' = \frac{1}{\sqrt{\alpha + x^2}}. \qquad (\arctan x)' = \frac{1}{1+x^2}.$$

To find derivatives of more complicated functions we use the following formulas:

$$(f + g)' = f' + g'. \qquad\qquad (fg)' = f'g + fg'.$$

$$(f - g)' = f' - g'. \qquad\qquad \left(\frac{f}{g}\right)' = \frac{gf' - fg'}{g^2}.$$

$$(cf)' = cf', \qquad c \text{ a constant.} \qquad \Big(f(g(x))\Big)' = f'(g(x))g'(x).$$

Now, some useful algebraic and trigonometric identities:

$$\alpha^x\alpha^y = \alpha^{x+y} \quad \text{and} \quad (\alpha^x)^y = \alpha^{xy}, \quad \text{where } \alpha > 0.$$

$$\log_a x = \frac{\ln x}{\ln a}, \quad \text{for } 0 < a \neq 1, \quad x > 0.$$

$$\ln xy = \ln x + \ln y, \quad \text{for } x > 0, \quad y > 0.$$

$$\ln\frac{x}{y} = \ln x - \ln y, \quad \text{for } x > 0, \quad y > 0.$$

$$\ln x^p = p\ln x, \quad \text{for } x > 0, \quad p \text{ an arbitrary number.}$$

$$\sin(x + y) = \sin x \cos y + \cos x \sin y.$$

$$\cos(x + y) = \cos x \cos y - \sin x \sin y.$$

$$\tan(x + y) = \frac{\tan x + \tan y}{1 - \tan x \tan y}.$$

$$(\cos x)^2 + (\sin x)^2 = 1.$$

$$\cos(-x) = \cos x. \qquad\qquad \sin(-x) = -\sin x.$$

$$\cos(x + \tfrac{\pi}{2}) = -\sin x. \qquad\quad \sin(x + \tfrac{\pi}{2}) = \cos x.$$

$$\cos(x + \pi) = -\cos x. \qquad\quad \sin(x + \pi) = -\sin x.$$

$$\cos(\pi - x) = -\cos x. \qquad\quad \sin(\pi - x) = \sin x.$$

$$\cos(x + 2\pi) = \cos x. \qquad\quad \sin(x + 2\pi) = \sin x$$

$$e^{\ln x} = x, \quad \text{for } x > 0. \qquad \ln e^x = x, \quad \text{for arbitrary } x.$$

$$\tan(\arctan x) = x, \quad \text{for arbitrary x.}$$

$$\arctan(\tan x) = x, \quad \text{for } -\tfrac{\pi}{2} \le x \le \tfrac{\pi}{2}.$$

$$\sin(\arcsin x) = x, \quad \text{for } -1 \le x \le 1.$$

$$\cos(\arccos x) = x, \quad \text{for } -1 \le x \le 1.$$

$$\arcsin(\sin x) = x, \quad \text{for } -\tfrac{\pi}{2} \le x \le \tfrac{\pi}{2}.$$

$$\arccos(\cos x) = x, \quad \text{for } 0 \le x \le \pi.$$

Finally, the most important power series expansions:

$$e^x = 1 + \frac{x}{1!} + \frac{x^2}{2!} + \frac{x^3}{3!} + \cdots = \sum_{k=0}^{\infty} \frac{x^k}{k!}.$$

$$\cos x = 1 - \frac{x^2}{2!} + \frac{x^4}{4!} - \frac{x^6}{6!} + \cdots = \sum_{k=0}^{\infty} (-1)^k \frac{x^{2k}}{(2k)!}.$$

$$\sin x = \frac{x}{1!} - \frac{x^3}{3!} + \frac{x^5}{5!} - \frac{x^7}{7!} + \cdots = \sum_{k=0}^{\infty} (-1)^k \frac{x^{2k+1}}{(2k+1)!}.$$

$$\ln(1+x) = \frac{x}{1} - \frac{x^2}{2} + \frac{x^3}{3} - \frac{x^4}{4} + \cdots = \sum_{k=0}^{\infty} \frac{(-1)^k}{k} x^k, \quad \text{for } -1 < x < 1.$$

$$\arctan x = \frac{x}{1} - \frac{x^3}{3} + \frac{x^5}{5} - \frac{x^7}{7} + \cdots = \sum_{k=0}^{\infty} (-1)^k \frac{x^{2k+1}}{2k+1}, \quad \text{for } -1 < x < 1.$$

$$(1+x)^\alpha = 1 + \binom{\alpha}{1} x + \binom{\alpha}{2} x^2 + \cdots = \sum_{k=0}^{\infty} \binom{\alpha}{k} x^k, \quad \text{for } -1 < x < 1.$$

$$(a+b)^m = \binom{m}{0} a^m + \binom{m}{1} a^{m-1} b + \binom{m}{2} a^{m-2} b^2 + \cdots + \binom{m}{m} b^m.$$

Most of the above formulas are presented in any calculus course. In this book they have been proved in a rigorous, analytic way.

CHAPTER 5

COMPLEMENTS TO THE THEORY OF LIMITS

5.1 IMPROPER LIMITS

By an *improper limit* we mean a limit the value of which is not a number but ∞ or $-\infty$.

5.1.1 Definition We say that a function f has at a the right-hand limit equal to ∞, and we write

$$f(a+) = \infty \quad \text{or} \quad \lim_{x \to a+} f(x) = \infty,$$

if for every number A there exists a number $\varepsilon > 0$ such that $f(x) > A$ for every x satisfying $a < x < a + \varepsilon$. Similarly, we say that a function f has at a the right-hand limit equal $-\infty$, and we write

$$f(a+) = -\infty \quad \text{or} \quad \lim_{x \to a+} f(x) = -\infty,$$

if for every number A there exists a number $\varepsilon > 0$ such that $f(x) < A$ for every x satisfying $a < x < a + \varepsilon$.

The left-hand limits are defined in a similar manner, we only need to replace inequalities $a < x < a + \varepsilon$ by $a - \varepsilon < x < a$ and the symbol $a+$ by $a-$.

For instance,

$$\lim_{x \to a+} \frac{1}{x - a} = \infty; \qquad \lim_{x \to a+} \frac{1}{a - x} = -\infty;$$

$$\lim_{x \to a-} \frac{1}{x - a} = -\infty; \qquad \lim_{x \to a-} \frac{1}{a - x} = \infty.$$

We can also speak of two-sided improper limits and write $\lim_{x \to a} f(x) = \infty$ whenever $f(a+) = f(a-) = \infty$, and similarly for $-\infty$.

Assume that $f(x) = 1/g(x) > 0$ holds on (a, b). Then the equalities $g(a+) = 0$ and $f(a+) = \infty$ are equivalent. This equivalence follows directly from the definition of improper limits. Similarly, equalities $g(b-) = 0$ and $f(b-) = \infty$ are equivalent. On the other hand, if $f(x) = 1/g(x) < 0$ on (a, b), then the equality $g(a+) = 0$ is equivalent to $f(a+) = -\infty$, and $g(b-) = 0$ is equivalent to $f(b-) = -\infty$.

We can also define improper limits at infinity.

5.1.2 Definition We say that a function f has the limit ∞ at infinity if for every number A there exists a number δ such that $f(x) > A$ for all $x > \delta$. We then write

(1) $$f(\infty) = \infty \quad \text{or} \quad \lim_{x \to \infty} f(x) = \infty.$$

Similarly, we say that a function f has the limit $-\infty$ at infinity if for every number A there exists a number δ such that $f(x) < A$ for all $x > \delta$. We then write

(2) $$f(\infty) = -\infty \quad \text{or} \quad \lim_{x \to \infty} f(x) = -\infty.$$

5.1.3 Examples

$$e^{\infty} = \infty, \qquad \ln \infty = \infty, \qquad \log_{\frac{1}{2}} \infty = -\infty,$$
$$\infty^{\alpha} = \infty \quad \text{for } \alpha > 0, \qquad \infty^{\alpha} = 0 \quad \text{for } \alpha < 0.$$

Note that (1) is equivalent to $\lim_{x \to 0+} f(1/x) = \infty$, and (2) to $\lim_{x \to 0+} f(1/x) = -\infty$. ∎

Exercises 5.1

1. Use Definition 5.1.1 to prove the following:

   (a) $\lim_{x \to 0+} \dfrac{1}{x} = \infty, \qquad \lim_{x \to 0-} \dfrac{1}{x} = -\infty.$

   (b) $\lim_{x \to 0+} \ln x = -\infty.$

   (c) $\lim_{x \to \frac{\pi}{2}-} \tan x = \infty, \qquad \lim_{x \to \frac{\pi}{2}+} \tan x = -\infty.$

2. Use Definition 5.1.2 to prove the limits in 5.1.3.

3. Prove that $\lim_{x \to \infty} f(x) = \infty$ if and only if $\lim_{x \to 0+} f(1/x) = \infty.$

5.2 THE EASY L'HOSPITAL THEOREM

When evaluating limits the so-called l'Hospital's rule (Guillaume Francois Antoine de l'Hospital, 1661–1704) can often be successfully used. It has two forms, an easy one and a difficult one.

5.2.1 Theorem (easy l'Hospital's rule) *Let f and g be differentiable on some interval $(a - \varepsilon, a + \varepsilon), \varepsilon > 0$. Then*

$$\lim_{x \to a} \frac{f(x)}{g(x)} = \lim_{x \to a} \frac{f'(x)}{g'(x)}$$

whenever $\lim_{x \to a} f(x) = \lim_{x \to a} g(x) = 0$, *and the limit* $\lim_{x \to a} \dfrac{f'(x)}{g'(x)}$ *(proper or improper) exists.*

This theorem is true for the left-hand limit, the right-hand limit, and the limit at infinity; that is, when a is replaced by $a-$, $a+$, or ∞, respectively. In the last case the interval is of the form (c, ∞).

Proof. The proof will be first carried out for the right hand limit. Let us then assume that the functions $f(x)/g(x)$ and $f'(x)/g'(x)$ are defined on an interval $(a, t) \subseteq (a, a + \varepsilon)$. We introduce an auxiliary function

$$\phi(x) = \begin{cases} f(t)g(x) - f(x)g(t) & \text{if } x \in (a, t] \\ 0 & \text{if } x = a. \end{cases}$$

Note that the function ϕ is continuous on $[a, t]$ and that $\phi(a) = \phi(t) = 0$. Since ϕ is also differentiable on $(a, a + \varepsilon)$ we have, by the mean value theorem, $\phi'(\xi) = 0$ for some $\xi \in (a, t)$. This means that

$$f(t)g'(\xi) - f'(\xi)g(t) = 0$$

or equivalently

$$\frac{f(t)}{g(t)} = \frac{f'(\xi)}{g'(\xi)}.$$

This implies that

$$\lim_{t \to a+} \frac{f(t)}{g(t)} = \lim_{\xi \to a+} \frac{f'(\xi)}{g'(\xi)},$$

whenever the second limit exists.

Replacing, in the above proof, the intervals $[a, t]$, (a, t), and $(a, a + \varepsilon)$ by $[t, a]$, (t, a), and $(a - \varepsilon, a)$, respectively, and the symbol $a+$ by $a-$ we obtain the proof for the left-hand limit. Hence, the theorem also holds for two-sided limits.

Finally, in the case $a = \infty$, we have

$$\lim_{x \to \infty} \frac{f(x)}{g(x)} = \lim_{x \to 0+} \frac{f(1/x)}{g(1/x)} = \lim_{x \to 0+} \frac{f'(1/x)(-1/x^2)}{g'(1/x)(-1/x^2)}$$

$$= \lim_{x \to 0+} \frac{f'(1/x)}{g'(1/x)} = \lim_{x \to \infty} \frac{f'(x)}{g'(x)},$$

and the proof is complete.

5.2.2 Examples

1. $\lim_{x \to 0} \dfrac{\sin x}{x} = \lim_{x \to 0} \dfrac{\cos x}{1} = 1,$

2. $\lim_{x \to 0} \dfrac{a^x - 1}{x} = \lim_{x \to 0} \dfrac{a^x \ln a}{1} = \ln a, \qquad a > 0$

3. $\lim_{x \to 0+} \dfrac{x}{1 - \cos x} = \lim_{x \to 0+} \dfrac{1}{\sin x} = \infty.$

Exercise 5.2

1. Find the following limits:

(a) $\lim_{x\to1} \dfrac{x^3 - 1}{x^3 - 2x^2 + 2x - 1}$;   (b) $\lim_{x\to a+} \dfrac{\sqrt{x^3 - a^3}}{\sqrt{x - a}}$;

(c) $\lim_{x\to0} \dfrac{a^x - b^x}{x}$; $a > 0, b > 0$;   (d) $\lim_{x\to0} \dfrac{e^x - e^{-x}}{\sin x}$;

(e) $\lim_{x\to0} \dfrac{\sin x - x \cos x}{(\sin x)^3}$;   (f) $\lim_{x\to0} \dfrac{e^x - e^{\sin x}}{x - \sin x}$.

5.3 THE DIFFICULT L'HOSPITAL THEOREM

In the second version of the l'Hospital theorem the assumptions that $\lim_{x\to a} f(x) = \lim_{x\to a} g(x) = 0$ is replaced by the assumption that $\lim_{x\to a} g(x) = \infty$.

5.3.1 Theorem (difficult l'Hospital's Rule). *Let f and g be differentiable on some interval $(a - \varepsilon, a + \varepsilon)$. Then*

$$\lim_{x\to a} \frac{f(x)}{g(x)} = \lim_{x\to a} \frac{f'(x)}{g'(x)}$$

whenever $\lim_{x\to a} g(x) = \infty$ and the limit $\lim_{x\to a} \dfrac{f'(x)}{g'(x)}$ (proper or improper) exists.

Proof. We will first prove the theorem under the assumption that $a = \infty$ and that $\lim_{x\to\infty} f'(x)/g'(x) = k$, where k is a finite number. Let ε be an arbitrary positive number. There exists a number x_0 such that

$$g(x) > 0 \quad \text{and} \quad \frac{f'(x)}{g'(x)} < k + \frac{1}{2}\varepsilon \quad \text{for} \quad x \geq x_0,$$

where $g'(x) \neq 0$. The derivative $g'(x)$ cannot change sign for $x \geq x_0$, and thus $g'(x) > 0$, because $g(x)$ tends to infinity as $x \to \infty$. Hence

(1) $$f'(x) - (k + \frac{1}{2}\varepsilon)g'(x) < 0 \quad \text{for} \quad x \geq x_0,$$

and furthermore

(2) $$f(x) - f(x_0) - (k + \frac{1}{2}\varepsilon)(g(x) - g(x_0)) < 0 \quad \text{for} \quad x > x_0,$$

because the left-hand side of (2) equals 0 at $x = x_0$ and has the left-hand side of (1) for its derivative. From (2) it follows that

$$\frac{f(x)}{g(x)} < k + \frac{1}{2}\varepsilon + \frac{f(x_0) - (k + \frac{1}{2}\varepsilon)g(x_0)}{g(x)} \quad \text{for} \quad x > x_0.$$

Since $g(\infty) = \infty$, there exists a number $x_1 \geq x_0$ such that

$$\frac{f(x_0) - (k + \frac{1}{2}\varepsilon)g(x_0)}{g(x)} < \frac{1}{2}\varepsilon \quad \text{for} \quad x \geq x_1.$$

Consequently,

(3)
$$\frac{f(x)}{g(x)} < k + \varepsilon \quad \text{for } x \geq x_1.$$

Similarly we can prove the existence of a number x_2 such that

(4)
$$\frac{f(x)}{g(x)} > k - \varepsilon \quad \text{for } x \geq x_2.$$

To this aim it suffices to change, in the above argument, the sign $<$ to $>$ and replace $\frac{1}{2}\varepsilon$ by $-\frac{1}{2}\varepsilon$.

From (3) and (4) we have $\lim_{x \to \infty} f(x)/g(x) = k$, proving the theorem in the special case.

Suppose now $\lim_{x \to \infty} f'(x)/g'(x) = \infty$. Then there exists a number x_0 such that

$$g(x) > 0 \quad \text{and} \quad \frac{f'(x)}{g'(x)} > 1 \quad \text{for } x \geq x_0.$$

Hence we have

$$f'(x) - g'(x) > 0 \quad \text{for } x \geq x_0,$$

and consequently

$$f(x) - f(x_0) - (g(x) - g(x_0)) > 0 \text{ for } x > x_0;$$

That is,

$$f(x) > g(x) - g(x_0) + f(x_0) \quad \text{for } x > x_0.$$

This shows that $\lim_{x \to \infty} f(x) = \infty$. Since $\lim_{x \to \infty} g'(x)/f'(x) = 0$ we have $\lim_{x \to \infty} g(x)/f(x) = 0$, due to the first part of this proof. This implies that $\lim_{x \to \infty} f(x)/g(x) = \infty$.

The case in which $\lim_{x \to \infty} f(x)/g(x) = -\infty$ can be reduced to the previous one by considering the quotient $-f(x)/g(x)$. In this way the theorem is proved for the limit at infinity. For the left- and right-hand limits at a point the assertion is obtained by considering the limits

$$\lim_{x \to a+} \frac{f(x)}{g(x)} = \lim_{x \to \infty} \frac{f(a + 1/x)}{g(a + 1/x)}$$

and

$$\lim_{x \to a-} \frac{f(x)}{g(x)} = \lim_{x \to \infty} \frac{f(a - 1/x)}{g(a - 1/x)}$$

and hence it follows for the two-sided limits.

5.3.2 Examples

1. $\lim_{x \to \infty} \dfrac{\ln x}{x^a} = \lim_{x \to \infty} \dfrac{1/x}{a x^{a-1}} = \lim_{x \to \infty} \dfrac{1}{a x^a} = 0 \quad \text{for} \quad a > 0.$

2. $\lim_{x \to \infty} \dfrac{e^x}{x} = \lim_{x \to \infty} \dfrac{e^x}{1} = \infty.$

3. More generally, for any real α we have $\lim_{x \to \infty} \dfrac{e^x}{x^\alpha} = \infty$, which is obvious for $\alpha \leq 0$. For $\alpha > 0$ we first note that

$$\frac{e^x}{x^\alpha} = \left(\frac{e^{x/\alpha}}{x}\right)^\alpha.$$

Hence we get

$$\lim_{x \to \infty} \frac{e^x}{x^\alpha} = \left(\lim_{x \to \infty} \frac{e^{x/\alpha}}{x}\right)^\alpha = \left(\lim_{x \to \infty} \frac{\frac{1}{\alpha}e^{x/\alpha}}{1}\right)^\alpha = \infty.$$

In this example we first use some auxiliary transformation and then l'Hospital's rule. A direct application of l'Hospital's rule leads to some difficulties.

4. Using induction we can also prove that

$$\lim_{x \to \infty} \frac{e^x}{x^n + a_{n-1}x^{n-1} + \ldots + a_1 x + a_o} = \infty,$$

for any $n \in N$ and $a_{n-1}, \ldots, a_0 \in R$. ∎

Exercises 5.3

1. Find the following limits:

    (a) $\lim_{x \to \infty} \dfrac{(\ln x)^\alpha}{x}$ (α arbitrary);

    (b) $\lim_{x \to 0+} \dfrac{\ln \tan x}{\ln \tan 2x}$;

    (c) $\lim_{x \to \infty} \dfrac{x^2}{2^x}$.

2. In the first part of the proof of Theorem 5.3.1 we claim that "the derivative $g'(x)$ cannot change sign for $x \geq x_0$." Justify that claim.

3. Prove that

    $$\lim_{x \to \infty} \frac{e^x}{x^n + a_{n-1}x^{n-1} + \ldots + a_1 x + a_0} = \infty$$

    for any $n \in N$ and any $a_{n-1}, \ldots, a_0 \in R$.

4. Justify the equality used in Example 5.2.3.3,

    $$\lim_{x \to \infty} \frac{e^x}{x^\alpha} = \left(\lim_{x \to \infty} \frac{e^{x/\alpha}}{x}\right)^\alpha.$$

5.4 SOME APPLICATIONS OF L'HOSPITAL'S RULE

If $\lim f(x) = 0$ and $\lim g(x) = \infty$ (the sign lim may denote any of the symbols $\lim_{x \to a+}$, $\lim_{x \to a-}$, $\lim_{x \to a}$, or $\lim_{x \to \infty}$, but the same for both functions), then $\lim f(x)g(x)$ can be computed by representing the product as

$$\frac{f(x)}{1/g(x)} \quad \text{or} \quad \frac{g(x)}{1/f(x)},$$

and then using the suitable form of l'Hospital's rule.

5.4.1 Examples

1. $$\lim_{x\to 0+} x(\ln x) = \lim_{x\to 0+} \frac{\ln x}{1/x} = \lim_{x\to 0+} \frac{1/x}{-1/x^2} = \lim_{x\to 0+} (-x) = 0.$$

2. $$\lim_{x\to\infty} x(\frac{1}{2}\pi - \arctan x) = \lim_{x\to\infty} \frac{\frac{1}{2}\pi - \arctan x}{1/x}$$

$$= \lim_{x\to\infty} \frac{-1/(1+x^2)}{-1/x^2} = \lim_{x\to\infty} \frac{-x^2}{1+x^2} = -1.$$

Similarly, if $\lim f(x) = \infty$ and $\lim g(x) = \infty$, then the limit of the difference $f - g$ can be found by representing that difference in the form

$$\frac{\dfrac{1}{g(x)} - \dfrac{1}{f(x)}}{\dfrac{1}{f(x)}\dfrac{1}{g(x)}},$$

and then using l'Hospital's rule.   ■

5.4.2 Examples

1. $$\lim_{x\to 0}\left(\frac{1}{x} - \frac{1}{\sin x}\right) = \lim_{x\to 0} \frac{\sin x - x}{x \sin x} = \lim_{x\to 0} \frac{\cos x - 1}{\sin x + x \cos x}$$

$$= \lim_{x\to 0} \frac{-\sin x}{2\cos x - x \sin x} = 0.$$

2. $$\lim_{x\to 1}\left(\frac{1}{\ln x} - \frac{1}{x-1}\right) = \lim_{x\to 1} \frac{(x-1) - \ln x}{(x-1)\ln x} = \lim_{x\to 1} \frac{1 - 1/x}{\ln x + (x-1)/x}$$

$$= \lim_{x\to 1} \frac{1/x^2}{1/x + 1/x^2} = \lim_{x\to 1} \frac{1}{x+1} = \frac{1}{2}.$$ ■

It often happens that it does not pay to use l'Hospital's rule, but is more convenient to use another method. For instance,

$$\lim_{x\to\infty} (\ln(1+x) - \ln x) = \lim_{x\to\infty} \ln\left(\frac{1}{x} + 1\right) = \ln\left(\lim_{x\to\infty}\left(\frac{1}{x} + 1\right)\right) = \ln 1 = 0.$$

We are going to discuss two other cases in which l'Hospital's rule gives good service.

If $f(x) > 0$, $\lim f(x) = 0$ and $\lim g(x) = 0$, then

(1) $$\lim f(x)^{g(x)} = \lim e^{g(x)\ln f(x)} = e^{\lim g(x)\ln f(x)}$$

and we apply l'Hospital's rule to the exponent. If the exponent w is complicated, then instead of e^w we often write $\exp(w)$. With this convention (1) can be written as

$$\lim f(x)^{g(x)} = \exp\left(\lim \frac{\ln f(x)}{1/g(x)}\right).$$

5.4.3 Example

$$\lim_{x \to 0+} x^x = \exp\left(\lim_{x \to 0+} \frac{\ln x}{1/x}\right) = \exp\left(\lim_{x \to 0+} \frac{1/x}{-1/x^2}\right) = \exp(0) = 1.$$

The following formula is useful

$$\lim (1 + f(x))^{1/g(x)} = \exp\left(\lim \frac{f'(x)}{g'(x)}\right), \text{ whenever } \lim f(x) = \lim g(x) = 0.$$

The sign lim may denote here, as well as in the preceding formulas, the left-hand, right-hand, or two-sided limit or the limit at infinity. The above formula is obtained from the following calculations

$$\lim (1 + f(x))^{1/g(x)} = \exp\left(\lim \frac{\ln(1 + f(x))}{g(x)}\right) = \exp\left(\lim \frac{\dfrac{f'(x)}{1 + f(x)}}{g'(x)}\right)$$

$$= \exp\left(\lim \frac{1}{1 + f(x)} \lim \frac{f'(x)}{g'(x)}\right) = \exp\left(1 \cdot \lim \frac{f'(x)}{g'(x)}\right). \quad\blacksquare$$

5.4.4 Examples

1. $\lim_{x \to 0}(1 + x)^{1/x} = \exp\left(\lim_{x \to 0} \frac{1}{1}\right) = e.$

2. $\lim_{x \to 0}(\cos x)^{1/x^2} = \lim_{x \to 0}(1 + (\cos x - 1))^{1/x^2}$

    $$= \exp\left(\lim_{x \to 0} \frac{-\sin x}{2x}\right) = \exp(-\frac{1}{2}) = \frac{1}{\sqrt{e}}.$$

3. $\displaystyle\lim_{x \to \infty}\left(1 - \frac{1}{x}\right)^x = \lim_{x \to 0+}(1 - x)^{1/x} = \exp\left(\lim_{x \to 0+} \frac{-1}{1}\right) = \exp(-1) = \frac{1}{e}.$ ■

Exercises 5.4

1. Find the following limits:

    (a) $\lim_{x \to 1}\left(\dfrac{x}{x - 1} - \dfrac{1}{\ln x}\right)$;   (b) $\lim_{x \to 0}\left(\dfrac{1}{x} - \dfrac{1}{x \tan x}\right)$;

    (c) $\lim_{x \to \pi/2}(x - \pi/2)\tan x$;   (d) $\lim_{x \to 0+} x(- \ln x)^\alpha$;

    (e) $\lim_{x \to 0+} x^{\sin x}$;   (f) $\lim_{x \to 0}(1 + mx)^{1/x}$;

    (g) $\lim_{x \to 1} x^{1/(1-x)}$,   (h) $\lim_{x \to 0}\left(\dfrac{\tan x}{x}\right)^{1/x^2}$.

2. Justify the equality

    $$\lim_{x \to \infty} \ln\left(\frac{1}{x} + 1\right) = \ln\left(\lim_{x \to \infty}\left(\frac{1}{x} + 1\right)\right).$$

5.5 FINDING LIMITS OF SEQUENCES

We know that if $\lim_{x \to \infty} f(x) = k$ and $a_n = f(n)$ for $n = 1, 2, 3, \ldots$, then $\lim_{n \to \infty} a_n = k$. Consequently, l'Hospital's rule can also serve in finding limits of sequences. For example, using this method it is easy to find the limit of the sequence $(1 + \frac{1}{n})^n$. In fact,

$$\lim_{x \to \infty} \left(1 + \frac{1}{x}\right)^x = \lim_{x \to 0+} (1 + x)^{1/x} = e.$$

We thus have the formula

$$\lim_{n \to \infty} \left(1 + \frac{1}{n}\right)^n = e.$$

This formula is often used as a definition of the number e. However, such a definition requires a proof of consistency: one has to prove that the limit exists. This can be accomplished by proving that the sequence is increasing and bounded, but the calculations are rather lengthy.

Changing to the limit of a function does not always work. For instance, it would be rather difficult to apply this method to the limit

$$\lim_{n \to \infty} \frac{1 + \frac{1}{2} + \cdots + \frac{1}{n}}{\ln n}.$$

In this and similar cases we can apply the following theorem, which is somewhat similar to the difficult l'Hospital's rule.

5.5.1 Theorem   *The equality*

$$\lim_{n \to \infty} \frac{a_n}{b_n} = \lim_{n \to \infty} \frac{a_{n+1} - a_n}{b_{n+1} - b_n}$$

holds whenever $b_{n+1} > b_n$ for all $n \in \mathbb{N}$, $\lim_{n \to \infty} b_n = \infty$, and the limit (proper or improper) on the right-hand side exists.

Proof.   Let us first assume that the limit

$$\lim_{n \to \infty} \frac{a_{n+1} - a_n}{b_{n+1} - b_n}$$

is finite and equal to α. Let ε be an arbitrary positive number. There exists an index n_0 such that

$$\alpha - \frac{\varepsilon}{2} < \frac{a_{n+1} - a_n}{b_{n+1} - b_n} < \alpha + \frac{\varepsilon}{2} \quad \text{for all} \quad n \geq n_0.$$

Hence

(1) $(\alpha - \frac{\varepsilon}{2})(b_{n+1} - b_n) < a_{n+1} - a_n < (\alpha + \frac{\varepsilon}{2})(b_{n+1} - b_n)$ for all $n \geq n_0.$

Adding inequalities (1) with indices $n = n_0, n_0 + 1, \ldots, p$ we get

$$(\alpha - \tfrac{\varepsilon}{2})(b_p - b_{n_0}) < a_p - a_{n_0} < (\alpha + \tfrac{\varepsilon}{2})(b_p - b_{n_0}).$$

Adding the term a_{n_0} and then dividing by b_p yields

$$\alpha - \frac{\varepsilon}{2} + \frac{a_{n_0} - (\alpha - \tfrac{1}{2}\varepsilon)b_{n_0}}{b_p} < \frac{a_p}{b_p} < \alpha + \frac{\varepsilon}{2} + \frac{a_{n_0} - (\alpha + \tfrac{\varepsilon}{2})b_{n_0}}{b_p}.$$

There exists an index p_0 such that for all $p > p_0$

$$\frac{a_{n_0} - (\alpha - \tfrac{1}{2}\varepsilon)b_{n_0}}{b_p} > -\frac{\varepsilon}{2} \quad \text{and} \quad \frac{a_{n_0} - (\alpha + \tfrac{1}{2}\varepsilon)b_{n_0}}{b_p} < \frac{\varepsilon}{2}.$$

Thus

$$\alpha - \varepsilon < \frac{a_p}{b_p} < \alpha + \varepsilon \quad \text{for all } p > p_0,$$

which proves that α is the limit of $\{a_p/b_p.\}$

If $\lim_{n\to\infty}(a_{n+1} - a_n)/(b_{n+1} - b_n) = \infty$, then $\lim_{n\to\infty} b_n = \infty$ implies that also $\lim_{n\to\infty} a_n = \infty$ and $a_{n+1} > a_n$ for all $n > n_0$. Thus, due to the case just considered, we have

$$\lim_{n\to\infty} \frac{b_n}{a_n} = \lim_{n\to\infty} \frac{b_{n+1} - b_n}{a_{n+1} - a_n} = 0$$

and, consequently, $\lim_{n\to\infty} a_n/b_n = \infty$. The case $\lim_{n\to\infty}(a_{n+1} - a_n)/(b_{n+1} - b_n) = -\infty$ reduces to the previous case by an easy substitution.

5.5.2 Examples

1.
$$\lim_{n\to\infty} \frac{1 + \tfrac{1}{2} + \cdots + \tfrac{1}{n}}{\ln n} = \lim_{n\to\infty} \frac{1/(n+1)}{\ln(n+1) - \ln n}$$

$$= \lim_{n\to\infty} \frac{n}{n+1} \frac{1}{n \ln(1 + 1/n)} = \lim_{n\to\infty} \frac{1}{\ln(1 + 1/n)^n} = \frac{1}{\ln e} = 1.$$

2.
$$\lim_{n\to\infty} \frac{\sqrt[n]{n!}}{n} = \lim_{n\to\infty} \exp\left(\frac{\ln n! - n \ln n}{n}\right)$$

$$= \exp\left(\lim_{n\to\infty} \frac{\ln(n+1)! - (n+1)\ln(n+1) - \ln n! + n \ln n}{1}\right)$$

$$= \exp\left(\lim_{n\to\infty}\left(-n \ln \frac{n+1}{n}\right)\right)$$

$$= \exp\left(\lim_{n\to\infty}(-\ln(1 + 1/n)^n)\right) = \exp(-\ln e) = \frac{1}{e}.$$

3.   If $\alpha > -1$, then

$$\lim_{n\to\infty} \frac{1^\alpha + 2^\alpha + \ldots + n^\alpha}{n^{\alpha+1}} = \lim_{n\to\infty} \frac{(n+1)^\alpha}{(n+1)^{\alpha+1} - n^{\alpha+1}}$$

$$= \lim_{n\to\infty} \left(\frac{n+1}{n}\right)^\alpha \lim_{n\to\infty} \frac{n^\alpha}{(n+1)^{\alpha+1} - n^{\alpha+1}}.$$

The first limit equals 1, whereas the second equals

$$\lim_{x\to 0+} \frac{\left(\dfrac{1}{x}\right)^\alpha}{\left(\dfrac{1}{x} + 1\right)^{\alpha+1} - \left(\dfrac{1}{x}\right)^{\alpha+1}} = \lim_{x\to 0+} \frac{x}{(1+x)^{\alpha+1} - 1}$$

$$= \lim_{x\to 0+} \frac{1}{(\alpha+1)(1+x)^\alpha} = \frac{1}{\alpha+1}.$$

Consequently,

$$\lim_{n\to\infty} \frac{1^\alpha + 2^\alpha + \cdots + n^\alpha}{n^{\alpha+1}} = \frac{1}{\alpha+1} \qquad (\alpha > -1).$$

Exercise 5.5

1.   Find the following limits:

(a) $\lim_{n\to\infty} \dfrac{\ln 1 + \ln 2 + \cdots + \ln n}{n \ln n}$,   (b) $\lim_{n\to\infty} \dfrac{\sqrt[n]{(n+1)(n+2)\cdots 2n}}{n}$.

5.6 D'ALEMBERT'S TEST

D'Alembert's test (also called the *ratio test*) is used to determine convergence of series. It is attributed to the French mathematician Jean le Rond d'Alembert (1717–1783).

5.6.1 Theorem (D'Alembert's test)   *The series*

(1)
$$\left\{ \sum_{k=1}^{n} a_k \right\}$$

is convergent if $\lim_{n\to\infty} |a_{n+1}/a_n| < 1$, *and divergent if* $\lim_{n\to\infty} |a_{n+1}/a_n| > 1$.

Proof.   Let $\lim_{n\to\infty} |a_{n+1}/a_n| = \alpha$. If $\alpha < 1$, then there exists a number q and an index n_0 such that

$$\alpha < q < 1 \quad \text{and} \quad \left|\frac{a_{n+1}}{a_n}\right| < q \quad \text{for} \quad n \geq n_0.$$

Thus we have the following inequalities:

$$|a_{n_0+1}| < |a_{n_0}|q,$$
$$|a_{n_0+2}| < |a_{n_0+1}|q < |a_{n_0}|q^2,$$

and generally, by induction,

(2)
$$|a_{n_0+n}| < |a_{n_0}|q^n.$$

Since the series $\left\{\sum_{k=0}^n a_{n_0} q^k\right\}$ is convergent, so is the series $\left\{\sum_{k=0}^n a_{n_0+k}\right\}$, by (2). This implies convergence of (1).

If $\alpha > 1$, then there is an index n_0 such that

$$\left|\frac{a_{n+1}}{a_n}\right| > 1 \quad \text{for} \quad n \geq n_0.$$

By induction we get

$$|a_{n_0+n}| > |a_{n_0}| > 0,$$

and hence the sequence does not converge to zero, implying that series (1) diverges.

5.6.2 Examples

1. For $a_n = n!/n^n$ we have

$$\left|\frac{a_{n+1}}{a_n}\right| = \frac{(n+1)!}{(n+1)^{n+1}} \cdot \frac{n^n}{n!} = \frac{1}{(1+1/n)^n}.$$

Since

$$\lim_{n\to\infty} \left|\frac{a_{n+1}}{a_n}\right| = \frac{1}{e} < 1,$$

the series

$$\left\{\sum_{k=1}^n \frac{k!}{k^k}\right\}$$

is convergent.

2. For $a_n = 3^n n!/n^n$, we have

$$\left|\frac{a_{n+1}}{a_n}\right| = \frac{3}{(1+1/n)^n} \to \frac{3}{e} > 1.$$

This proves that the series

$$\left\{\sum_{k=1}^n \frac{3^k k!}{k^k}\right\}$$

is divergent.

Exercises 5.6

1. Give an example of a convergent series $\left\{ \sum_{k=1}^{n} a_k \right\}$ for which $\lim_{n\to\infty} \left| \frac{a_{n+1}}{a_n} \right| = 1$.

2. Give an example of a convergent series $\left\{ \sum_{k=1}^{n} a_k \right\}$ for which $\lim_{n\to\infty} \left| \frac{a_{n+1}}{a_n} \right| = 1$.

3. Prove *Cauchy's test* (called also the *root test*):
   The series $\left\{ \sum_{k=1}^{n} a_k \right\}$ is convergent if $\lim_{n\to\infty} |a_n|^{1/n} < 1$, and divergent if $\lim_{n\to\infty} |a_n|^{1/n} > 1$.

4. Test the following for convergence:

   (a) $\left\{ \sum_{k=1}^{n} \dfrac{k^3}{(-e)^k} \right\}$; (b) $\left\{ \sum_{k=1}^{n} \dfrac{(k!)^2}{(2k)!} \right\}$;

   (c) $\left\{ \sum_{k=1}^{n} \dfrac{k!}{2^k + 1} \right\}$;(d) $\left\{ \sum_{k=1}^{n} \dfrac{7k + k^7 + 7^k}{k! + \ln(k + 7)} \right\}$;

   (e) $\left\{ \sum_{k=1}^{n} \dfrac{\sin \pi^k}{\pi^k} \right\}$; (f) $\left\{ \sum_{k=1}^{n} \dfrac{2^k k!}{k^k} \right\}$.

5.7 THEOREM OF ABEL

It may happen that a series $\left\{ \sum_{k=1}^{n} a_k \right\}$ is convergent, but the series $\left\{ \sum_{k=1}^{n} |a_k| \right\}$ is not. Such a series is called *conditionally convergent*. It is clear that conditional convergence can never be proved by the comparison test (Theorem 4.6.1). Leibniz's theorem (4.36.1), which says that if $\{a_n\}$ is monotone and convergent to zero the series $\left\{ \sum_{k=1}^{n} (-1)^{k+1} a_k \right\}$ is convergent, can be often successfully used. For example, the series

(1) $$\left\{ \sum_{k=1}^{n} \frac{(-1)^{k+1}}{k} \right\} \quad \text{and} \quad \left\{ \sum_{k=1}^{n} \frac{(-1)^{k+1}}{2k - 1} \right\}$$

are convergent, by Leibniz's theorem. That theorem, however, does not help in finding the sums of series (1). Evaluation of those sums is much more difficult than proving convergence. It can be done by using the following theorem due to the Norwegian mathematician Niels Henrik Abel (1802–1829).

5.7.1 Theorem   *If the series $\left\{ \sum_{k=0}^{n} a_k \right\}$ is convergent, then the function*

$$f(x) = \sum_{k=0}^{\infty} a_k x^k$$

is left-hand continuous at $x = 1$; that is,

$$\lim_{x\to 1-} f(x) = \sum_{k=0}^{\infty} a_k.$$

Proof. Since the sequence of partial sums $s_n = a_0 + \ldots + a_n$ is convergent, it is bounded. Hence the series $\left\{ \sum_{k=0}^{n} s_k x^k \right\}$ is convergent for $|x| < 1$. Moreover,

$$(1 - x) \sum_{k=0}^{\infty} s_k x^k = \sum_{k=0}^{\infty} s_k x^k - \sum_{k=0}^{\infty} s_k x^{k+1} = \sum_{k=0}^{\infty} a_k x^k,$$

and thus

(2)
$$f(x) = (1 - x) \sum_{k=0}^{\infty} s_k x^k.$$

Let $s = \sum_{k=0}^{\infty} a_k$. Since $1/(1 - x) = \sum_{k=0}^{\infty} x^k$, we have

(3)
$$s = (1 - x) \sum_{k=0}^{\infty} s x^k.$$

From (2) and (3) we get

(4)
$$f(x) - s = (1 - x) \sum_{k=0}^{\infty} (s_k - s) x^k.$$

Let ε be an arbitrary positive number. Since $\lim_{k \to \infty} s_k = s$, there exists an index m such that $|s_k - s| < \varepsilon/2$ for all $k > m$. Equality (4) can be rewritten in the form

$$f(x) - s = (1 - x) \left(\sum_{k=0}^{m} (s_k - s) x^k + \sum_{k=m+1}^{\infty} (s_k - s) x^k \right).$$

Letting $p = \sum_{k=0}^{m} |s_k - s|$ we have, for $0 < x < 1$,

$$|f(x) - s| < (1 - x) p + (1 - x) \sum_{k=m+1}^{\infty} \frac{\varepsilon}{2} x^k < (1 - x) p + \frac{\varepsilon}{2}.$$

There is a number $\delta > 0$ such that $(1 - x) p < \varepsilon/2$ for $1 - \delta < x < 1$. Thus

$$|f(x) - s| < \varepsilon \quad \text{for} \quad 1 - \delta < x < 1,$$

which proves that $\lim_{x \to 1-} f(x) = s$.

In Section 4.24 we found that

(5) $\quad \ln(1 + x) = \dfrac{x}{1} - \dfrac{x^2}{2} + \dfrac{x^3}{3} - \dfrac{x^4}{4} + \cdots = \displaystyle\sum_{k=1}^{\infty} \dfrac{(-1)^{k+1} x^k}{k} \quad$ for $\quad |x| < 1$.

In view of Abel's theorem the equality must also hold for $x = 1$, since the function $\ln(1 + x)$ is continuous at $x = 1$. Thus

$$\ln 2 = \frac{1}{1} - \frac{1}{2} + \frac{1}{3} - \frac{1}{4} + \cdots.$$

Similarly, from the equation

$$\arctan x = \frac{x}{1} - \frac{x^3}{3} + \frac{x^5}{5} - \frac{x^7}{7} + \cdots = \sum_{k=1}^{\infty} \frac{(-1)^{k+1} x^{2k-1}}{2k-1} \quad \text{for} \quad |x| < 1$$

it follows by Abel's theorem that

$$\frac{\pi}{4} = \frac{1}{1} - \frac{1}{3} + \frac{1}{5} - \frac{1}{7} + \cdots .$$

These series, known already by Leibniz, have a very elegant form but they cannot be used to calculate approximate values of $\ln 2$ or π. For example, if we wanted to compute π with 5-decimal-place accuracy, we would have to use about 100,000 terms of the sequence and evaluate each of them with 10-decimal-place accuracy.

Exercises 5.7

1. Prove that

$$\frac{1}{2 \cdot 1!} - \frac{1}{2^2 \cdot 2!} + \frac{1 \cdot 3}{2^3 \cdot 3!} - \frac{1 \cdot 3 \cdot 5}{2^4 \cdot 4!} + \cdots = \sqrt{2} - 1.$$

2. Prove the following version of Abel's theorem:

   If $\left\{\sum_{k=0}^{n} a_k\right\}$ converges, then $\left\{\sum_{k=0}^{n} a_k x^k\right\}$ converges uniformly on $[0, 1]$.

CHAPTER 6

ANTIDERIVATIVES

6.1 ANTIDERIVATIVES

We say that F is an *antiderivative* of f on an interval (finite or infinite), if for every x in that interval we have

$$(1) \qquad\qquad F'(x) = f(x).$$

For example, $\sin x$ is an antiderivative of $\cos x$ on $(-\infty, \infty)$ because $(\sin x)' = \cos x$.

If F is an antiderivative of f, then

(a) $F + c$ is an antiderivative of f for any constant $c \in \mathbb{R}$;

(b)   Each antiderivative of f is of the form $F + c$, since functions with the same derivative differ by a constant.

Therefore, it suffices to know a single antiderivative of f in order to obtain all of them by adding a constant. We often say that antiderivatives are determined up to an additive constant.

Antiderivatives are also called *indefinite integrals* and denoted by

$$\int f(x)dx.$$

The word "indefinite" refers to the fact that this symbol does not denote any specific function, but one of an infinite collection of functions differing from one another by an additive constant.

The calculation of antiderivatives, called *integration*, is the inverse of differentiation. For this reason we can use formulas derived for differentiation to obtain formulas for antiderivatives. We provide in the following table some antiderivatives. For purely practical purposes we supply these formulas in slightly modified form, adjusted to convenient calculations. The formulas are to be understood such that the function on the right represents one of the possible antiderivatives of the function on the left.

$f(x)$		$\int f(x)\,dx$
c	(constant)	cx
x^a	$(a \neq -1)$	$\dfrac{x^{a+1}}{a+1}$
$\dfrac{1}{x}$		$\ln|x|$
e^x		e^x
a^x	$(a > 0 \quad \text{and} \quad a \neq 1)$	$\dfrac{a^x}{\ln a}$
$\dfrac{1}{a^2 + x^2}$	$(a > 0)$	$\dfrac{1}{a}\arctan\dfrac{x}{a}$
$\dfrac{1}{\sqrt{a^2 - x^2}}$	$(a > 0)$	$\arcsin\dfrac{x}{a}$
$\dfrac{1}{\sqrt{a + x^2}}$	$(a > 0)$	$\ln\left(x + \sqrt{a + x^2}\right)$
$\cos x$		$\sin x$
$\sin x$		$-\cos x$
$\dfrac{1}{(\cos x)^2}$		$\tan x$
$\dfrac{1}{(\sin x)^2}$		$-\dfrac{\cos x}{\sin x}$
$\dfrac{1}{\sin x}$		$\ln\left|\tan\dfrac{x}{2}\right|$

Verification of these formulas consists of differentiation of the functions on the right-hand side, yielding the functions on the left-hand side.

Integration is usually much more difficult than differentiation. For instance, $\int \sin x^2\,dx$ cannot be calculated at all. This does not mean that the antiderivative of $\sin x^2$ does not exist. In Section 7.15 it will be proved that every continuous function has an antiderivative and so, in particular, there exists a function f such that $f'(x) = \sin x^2$. But that function cannot be expressed in terms of elementary functions nor by any finite combination of them (meaning by elementary function rational functions, radicals, exponential functions, logarithms, and trigonometric and cyclometric (inverse trigonometric) functions). Sometimes the integral of a simple function, although expressible by elementary functions alone, may be rather complicated. For example,

$$\int \frac{dx}{1 + x^4} = \frac{1}{4\sqrt{2}}\ln\frac{x^2 + x\sqrt{2} + 1}{x^2 - x\sqrt{2} + 1} + \frac{1}{2\sqrt{2}}\arctan\frac{x\sqrt{2}}{1 - x^2}$$

(see the remarks at the end of Section 6.4). It is therefore necessary to develop some general methods of integration, which are the subject of this chapter.

In this chapter, instead of "indefinite integral" we shall say simply "integral." In the next chapter we shall define a different type of integral and then the distinction will be necessary.

6.2 INTEGRATION BY SUBSTITUTION

We have the following simple formulas:

$$\int (f(x) + g(x))dx = \int f(x)dx + \int g(x)dx,$$

$$\int (f(x) - g(x))dx = \int f(x)dx - \int g(x)dx,$$

$$\int cf(x)dx = c\int f(x)dx.$$

They can be expressed in words as follows. *The integral of the sum equals the sum of the integrals. The integral of the difference equals the difference of the integrals. The integral of a function multiplied by a constant equals that constant times the integral of the function.*

The equalities in the above (and all similar) formulas should be understood to hold to within an additive constant. They can easily be proved by differentiation of both sides. In fact, from the definition of the indefinite integral it follows that

$$\left(\int f(x)dx\right)' = f(x).$$

6.2.1 Examples

1. $\int \left(x^2 + \dfrac{1}{3x}\right)dx = \int x^2 dx + \dfrac{1}{3}\int \dfrac{dx}{x} = \dfrac{x^3}{3} + \dfrac{1}{3}\ln|x|.$

2. $\int \left(e^x - 1\right)dx = \int e^x dx - \int dx = e^x - x.$ ■

In evaluating integrals we often apply the formula

(1) $F(g(t)) = \int f(g(t))g'(t)dt \quad$ where $F(t) = \int f(x)dx,$

which is the basis for the method called *integration by substitution*.

As before, to prove the formula it suffices to differentiate both sides. Formula (1) is often written as

(2) $\int f(x)\,dx = \int f(g(t))\,g'(t)\,dt \quad$ where $x = g(t).$

In applications we usually interpret the formula in a different way. We regard both sides of (2) as functions of x. Then, on the right-hand side, which is actually a function of t, we change the variable to x using the substitution $t = h(x)$, where h is the inverse of g. This can be done if g is invertible. The following example will explain it better.

We want to evaluate the integral

$$\int \frac{dx}{ax + b} \qquad (a \neq 0).$$

We set $ax + b = t$ or, equivalently, $x = (t - b)/a$. Then on the right-hand side of (1) we have

$$\int \frac{1}{t}\frac{1}{a}\, dt = \frac{1}{a}\int \frac{dt}{t} = \frac{1}{a}\ln|t|.$$

Changing back to the variable x we get

$$\int \frac{dx}{ax + b} = \frac{1}{a}\ln|ax + b| \qquad (a \neq 0).$$

The substitution $ax + b = t$ is one of the substitutions most frequently used in integrating. We should develop the ability to write down the final result of this substitution without any additional calculations. Let us evaluate, for example, the integral

$$\int \frac{dx}{1 + x + x^2}.$$

We first write it in the form

$$\int \frac{dx}{3/4 + \left(x + 1/2\right)^2}.$$

Then, after making, mentally, the substitution $x + \frac{1}{2} = t$, we write the result

$$\frac{1}{\sqrt{3/4}}\arctan \frac{x + 1/2}{\sqrt{3/4}},$$

obtaining, finally,

$$\int \frac{dx}{1 + x + x^2} = \frac{2}{\sqrt{3}}\arctan \frac{2x + 1}{\sqrt{3}}.$$

Exercises 6.2

1. Evaluate the following integrals:

(a) $\int 3 \cdot 2^x + 2 \cdot 3^x \, dx$;

(b) $\int a \sin x + b \cos x \, dx$;

(c) $\int \frac{1}{1 + x^2} + \frac{1}{2 + x^2} + \frac{1}{3 + x^2}\, dx$;

(d) $\int \frac{1}{\sqrt{1 + x^2}} - \frac{1}{\sqrt{1 - x^2}}\, dx$;

(e) $\int e^{ax+b}\, dx$;

(f) $\int \cos 2x \, dx$;

(g) $\int \sin(3x + 1)\, dx$;

(h) $\int \dfrac{dx}{1 - x + 2x^2}$;

(i) $\int \dfrac{dx}{\sqrt{1 - x + 2x^2}}$;

(j) $\int \dfrac{dx}{\sqrt{1 - x - 2x^2}}$.

6.3 INTEGRATION BY PARTS

We often use *integration by parts,* which consists of using the formula

(1) $$\int f(x)g'(x)\, dx = f(x)g(x) - \int f'(x)g(x)\, dx,$$

which is proved by differentiation of both sides.

6.3.1 Examples

1. In order to find $\int x \cos x\, dx$ we write

$$f(x) = x, \qquad g'(x) = \cos x,$$
$$f'(x) = 1, \qquad g(x) = \sin x.$$

In view of (1) we have

$$\int x \cos x\, dx = x \sin x - \int \sin x\, dx = x \sin x + \cos x.$$

2. To find $\int \ln x\, dx$ we write

$$f(x) = \ln x, \qquad g'(x) = 1,$$
$$f'(x) = \frac{1}{x}, \qquad g(x) = x,$$

and by (1)

$$\int \ln x\, dx = x \ln x - \int \frac{1}{x} x\, dx = x \ln x - x. \qquad \blacksquare$$

Systematic use of the scheme

$$\begin{matrix} f & g' \\ f' & g \end{matrix}$$

facilitates the work and helps in avoiding errors. In the first line of the table stand the factors of the integrand; on the main diagonal we have the factors already integrated; and in the bottom line we find the factors to be put under the integral sign on the right-hand side of the equation. Using this scheme we do not even have to write the letters f and g.

6.3.2 Examples

1. To calculate the integral $\int x^2 e^x \, dx$ we write

$$
\begin{matrix} x^2 & e^x \\ 2x & e^x \end{matrix} \qquad \int x^2 e^x \, dx = x^2 e^x - 2 \int x e^x \, dx.
$$

To find the remaining integral we write

$$
\begin{matrix} x & e^x \\ 1 & e^x \end{matrix} \qquad \int x e^x \, dx = x e^x - \int e^x \, dx = x e^x - e^x.
$$

Thus we have

$$
\int x^2 e^x \, dx = (x^2 - 2x + 2) e^x.
$$

2. For the integral $I = \int \dfrac{\ln x}{x} dx$ we write

$$
\begin{matrix} \ln x & 1/x \\ 1/x & \ln x \end{matrix} \qquad I = (\ln x)^2 - \int \frac{\ln x}{x} \, dx = (\ln x)^2 - I.
$$

Hence $I = \frac{1}{2}(\ln x)^2$.

Exercise 6.3 ────────────────────────────────

1. Integrate the following functions:

   (a) $\int \arctan x \, dx$;

   (b) $\int \arcsin x \, dx$;

   (c) $\int \arccos x \, dx$;

   (d) $\int \dfrac{\arctan x}{1 + x^2} \, dx$;

   (e) $\int \dfrac{\arcsin x}{\sqrt{1 - x^2}} \, dx$;

   (f) $\int x^3 \sin x \, dx$;

   (g) $\int e^x \cos x \, dx$;

   (h) $\int \arctan (2x - 1) \, dx$;

   (i) $\int x e^{2x} \sin \frac{1}{2} x \, dx$.

6.4 INTEGRATION OF RATIONAL FUNCTIONS

We shall now discuss techniques for the integration of rational functions.

6.4.1 Example

When dealing with the integral $\int \dfrac{dx}{(1 - x^2)}$ we first decompose the integrand into simple fractions:

$$
\frac{1}{1 - x^2} = \frac{1}{2}\left(\frac{1}{1 + x} + \frac{1}{1 - x} \right).
$$

Hence

$$\int \frac{dx}{1-x^2} = \frac{1}{2}\int \frac{dx}{1+x} + \frac{1}{2}\int \frac{dx}{1-x} = \frac{1}{2}\ln|1+x| - \frac{1}{2}\ln|1-x| = \frac{1}{2}\ln\left|\frac{1+x}{1-x}\right|.$$

■

6.4.2 Example   To evaluate the integral $\int \dfrac{x^2 dx}{x^2 - 2x - 3}$ we write

$$\frac{x^2}{x^2 - 2x - 3} = 1 + \frac{2x + 3}{x^2 - 2x - 3} = 1 + \frac{2x + 3}{(x-3)(x+1)} = 1 + \frac{A}{x-3} + \frac{B}{x+1}.$$

We find numbers A and B that satisfy the last equality or, which turns out to be the same, satisfy the equation

$$2x + 3 = A(x + 1) + B(x - 3).$$

Comparing coefficients it is easy to find that $A = \frac{9}{4}$ and $B = -\frac{1}{4}$. Thus

$$\int \frac{x^2 dx}{x^2 - 2x - 3} = \int dx + \frac{9}{4}\int \frac{dx}{x-3} - \frac{1}{4}\int \frac{dx}{x+1} = x + \frac{9}{4}\ln|x-3| - \frac{1}{4}\ln|x+1|.$$

■

6.4.3 Example
We have to evaluate the integral

$$\int \frac{x^3 dx}{x^2 - 2x + 3}$$

We write

$$\frac{x^3}{x^2 - 2x + 3} = x + 2 + \frac{x - 6}{x^2 - 2x + 3}.$$

■

—And I presume that the last fraction should be decomposed into a simple fractions as in the preceding examples.

—No, it cannot be done.

—Why not?

—Because $x^2 - 2x + 3$ cannot be decomposed into linear factors. It depends on the discriminant. By the discriminant of $Ax^2 + Bx + C$ we mean $\Delta = B^2 - 4AC$. If $\Delta > 0$, then decomposition into two factors is possible:

$$Ax^2 + Bx + C = A\left(x - \frac{-B - \sqrt{\Delta}}{2A}\right)\left(x - \frac{-B + \sqrt{\Delta}}{2A}\right);$$

if $\Delta < 0$ then we make the following transformation:

$$\frac{Dx + E}{Ax^2 + Bx + C} = \frac{D}{2A}\frac{(Ax^2 + Bx + C)'}{(Ax^2 + Bx + C)} + \frac{2AE - BD}{2A^2}\frac{1}{\left(x + B/2A\right)^2 + -\Delta/4A^2}.$$

—This formula is very complicated—almost impossible to remember.

—It would be useless to remember it. We should rather remember the general method and not the details. In our example, we have

$$\frac{x - 6}{x^2 - 2x + 3} = \frac{1}{2}\frac{2x - 2}{x^2 - 2x + 3} - \frac{5}{(x - 1)^2 + 2}$$

and hence

$$\int \frac{x - 6}{x^2 - 2x + 3}\, dx = \frac{1}{2}\ln(x^2 - 2x + 3) - \frac{5}{\sqrt{2}}\arctan\frac{x - 1}{\sqrt{2}}.$$

6.4.4 Example

$$\int \frac{dx}{x^3 + 2x^2 + 3x + 6}.$$

Here the integrand is not a simple fraction. By *simple fractions* we mean expressions of the form

$$\frac{1}{(Ax + B)^n} \quad \text{or} \quad \frac{Dx + E}{(Ax^2 + Bx + C)^n}$$

where $n \in \mathbb{N}$. Since every polynomial can be decomposed into a product of polynomials of the first and the second degree, every rational function can be represented as a sum of simple fractions. In our case we have

$$x^3 + 2x^2 + 3x + 6 = (x + 2)(x^2 + 3).$$

Hence

$$\frac{1}{x^3 + 2x^2 + 3x + 6} = \frac{A}{x + 2} + \frac{Bx + C}{x^2 + 3},$$

and, from the equation

$$1 = A(x^2 + 3) + (Bx + C)(x + 2),$$

we find

$$A = 1/7, \qquad B = -1/7, \qquad C = 2/7.$$

Hence

$$\int \frac{dx}{x^3 + 2x^2 + 3x + 6} = \frac{1}{7}\int \frac{dx}{x + 2} - \frac{1}{14}\int \frac{2x\, dx}{x^2 + 3} + \frac{2}{7}\int \frac{dx}{x^2 + 3}$$

$$= \frac{1}{7}\ln|x + 2| - \frac{1}{14}\ln|x^2 + 3| + \frac{2}{7\sqrt{3}}\arctan\frac{x}{\sqrt{3}}. \quad \blacksquare$$

Therefore, to integrate a rational function we have to be able to decompose its denominator and then we need to know how to evaluate the following integrals:

$$\int \frac{dx}{(Ax + B)^n} \quad \text{and} \quad \int \frac{Dx + E}{(Ax^2 + Bx + C)^n}, \qquad n \in \mathbb{N}.$$

The first integral is easy:

$$\int \frac{dx}{(Ax + B)^n} = \frac{1}{A(1 - n)(Ax + B)^{n-1}}.$$

Thus it remains to consider the second integral. The assumption that the denominator does not decompose into linear factors implies $\Delta = B^2 - 4AC < 0$. Thus, using the substitution $x + B/(2A) = t$, we obtain after a few simple algebraic calculations

$$\frac{Dx + E}{(Ax^2 + Bx + C)^n} = \alpha \frac{2t}{(t^2 + r^2)^n} + \beta \frac{1}{(t^2 + r^2)^n} \quad n \in \mathbb{N},$$

where $r^2 = -\Delta/(4A^2)$ and the coefficients α and β are properly chosen. Now, the first fraction is the derivative of

$$\frac{1}{(1 - n)(t^2 + r^2)^{n-1}}$$

so that

$$\int \frac{2t\,dt}{(t^2 + r^2)^n} = \frac{1}{(1 - n)(t^2 + r^2)^{n-1}}.$$

We still have to find the integral

$$(1) \qquad J_n = \int \frac{dt}{(t^2 + r^2)^n}.$$

We shall show that

$$(2) \qquad J_{n+1} = \frac{1}{2nr^2} \frac{t}{(t^2 + r^2)^n} + \frac{2n - 1}{2n} \frac{1}{r^2} J_n.$$

This formula allows us to reduce the degree of the power in the denominator by 1 and after n steps we have to evaluate only J_1:

$$J_1 = \int \frac{dt}{t^2 + r^2} = \frac{1}{r} \arctan \frac{t}{r}.$$

Applying integration by parts to (1) we write

$$\begin{array}{cc} \dfrac{1}{(t^2 + r^2)^n} & 1 \\[2em] -\dfrac{2nt}{(t^2 + r^2)^{n+1}} & t \end{array}$$

and

$$(3) \qquad J_n = \frac{t}{(t^2 + r^2)^n} + 2n \int \frac{t^2\,dt}{(t^2 + r^2)^{n+1}}.$$

The last integral may be transformed in the following way:

$$\int \frac{t^2\,dt}{(t^2 + r^2)^{n+1}} = \int \frac{(t^2 + r^2) - r^2}{(t^2 + r^2)^{n+1}}\,dt$$

$$= \int \frac{dt}{(t^2 + r^2)^n} - r^2 \int \frac{dt}{(t^2 + r^2)^{n+1}} = J_n - r^2 J_{n+1}.$$

Substituting this in (3) we get

$$J_n = \frac{t}{(t^2 + r^2)^n} + 2nJ_n - 2nr^2 J_{n+1}$$

whence (2) follows.

It is interesting that, when applying a general method, a simple example can lead to complicated calculations. Consider, for example,

$$\int \frac{x\,dx}{x^4 + 1}.$$

Since

$$x^4 + 1 = (x^2 - x\sqrt{2} + 1)(x^2 + x\sqrt{2} + 1),$$

we look for decomposition of the form

$$\frac{x}{x^4 + 1} = \frac{Ax + B}{x^2 - x\sqrt{2} + 1} + \frac{Cx + D}{x^2 + x\sqrt{2} + 1}.$$

Comparing coefficients in the equation

$$x = (Ax + B)(x^2 + x\sqrt{2} + 1) + (Cx + D)(x^2 - x\sqrt{2} + 1)$$

we find $A = 0$, $B = 1/(2\sqrt{2})$, $C = 0$, and $D = -1/(2\sqrt{2})$. Thus

$$\frac{x}{x^4 + 1} = \frac{1}{2\sqrt{2}}\left(\frac{1}{x^2 - x\sqrt{2} + 1} - \frac{1}{x^2 + x\sqrt{2} + 1}\right)$$

and finally

$$\int \frac{x\,dx}{x^4 + 1} = \frac{1}{2}\left(\arctan(x\sqrt{2} - 1) - \arctan(x\sqrt{2} + 1)\right).$$

—I can calculate this integral in a much simpler way: I substitute $x^2 = t$ and obtain

$$\int \frac{x\,dx}{x^4 + 1} = \frac{1}{2}\int \frac{dt}{t^2 + 1} = \frac{1}{2}\arctan t = \frac{1}{2}\arctan x^2.$$

—Indeed. Your work is more like the work of an artist, whereas the use of standard methods is like the work of a craftsman. In integration we should always look for special tricks and only when they do not work, apply general methods.

—And how do you explain that we have obtained different results?

—They only look different. In fact one can check, using the formula in Section 4.35, that they differ by an additive constant.

—The above trick does not work if we erase the x in the numerator. What do we do in this case, $\int \dfrac{dx}{x^4+1}$?

—It is said that Leibniz could not evaluate this integral, because he did not know how to factor the polynomial x^4+1. Knowing that decomposition and using the general method one can obtain

$$\int \frac{dx}{x^4+1} = \frac{1}{4\sqrt{2}} \ln\left|\frac{x^2+x\sqrt{2}+1}{x^2-x\sqrt{2}+1}\right| + \frac{1}{2\sqrt{2}} \arctan \frac{x\sqrt{2}}{1-x^2}.$$

Exercises 6.4

1. Integrate the following functions:

   (a) $\int \dfrac{x^4}{x^4-1}\,dx$;

   (b) $\int \dfrac{dx}{x^3+1}$;

   (c) $\int \dfrac{dx}{x^4+x^2+1}$;

   (d) $\int \dfrac{dx}{(1+x^2)^2}$;

   (e) $\int \dfrac{3x+5}{(x^2+2x+2)^2}\,dx$;

   (f) $\int \dfrac{dx}{x(x^7+1)}$;

   (g) $\int \dfrac{dx}{x(x^5+1)^2}$;

   (h) $\int \dfrac{x^2\,dx}{(x-1)^{10}}$.

2. Prove that $\arctan(x\sqrt{2}-1) - \arctan(x\sqrt{2}+1) = \arctan x^2 - \frac{\pi}{2}$.

6.5 INTEGRATION OF SOME IRRATIONAL EXPRESSIONS

The integral of a rational function of x and $\sqrt{ax^2+bx+c}$ can always be expressed by elementary functions. This theorem is due to Leonhard Euler (1707–1783) who introduced three types of substitutions, called Euler's substitutions. However, that general theory is seldom applied in practice, because with a little invention one can often find a better way to obtain the result.

6.5.1 Examples

1. Find the integral $I = \int \dfrac{dx}{\sqrt{x^2-6x+15}}$. We substitute $x-3=t$ and obtain

$$I = \int \frac{dt}{\sqrt{t^2+6}} = \ln\left(t+\sqrt{t^2+6}\right) = \ln\left(x-3+\sqrt{x^2+6x+15}\right).$$

2. In the integral $I = \int \dfrac{dx}{\sqrt{4-2x-x^2}}$ we substitute $x+1=\sqrt{5}t$ and get

$$I = \int \frac{dt}{\sqrt{1-t^2}} = \arcsin t = \arcsin \frac{x+1}{\sqrt{5}}.$$

3. $I = \int \dfrac{x-5}{\sqrt{x^2-2x+5}}\,dx = \dfrac{1}{2}\int \dfrac{2x-2}{\sqrt{x^2-2x+5}}\,dx - 4\int \dfrac{dx}{\sqrt{x^2-2x+5}}$

$\qquad = \sqrt{x^2-2x+5} - 4\int \dfrac{dx}{\sqrt{(x-1)^2+4}}$

$\qquad = \sqrt{x^2-2x+5} - 4\ln\left| x-1 + \sqrt{x^2-2x+5}\right|.$

4. To evaluate the following integral we start by integrating by parts

$$I = \int \sqrt{x^2-2x+5}\,dx = x\sqrt{x^2-2x+5} - I_1,$$

where

$$I_1 = \int \frac{x^2-x}{\sqrt{x^2-2x+5}}\,dx = I - \int \frac{x-5}{\sqrt{x^2-2x+5}}\,dx.$$

Hence

$$I = \frac{1}{2}\left(x\sqrt{x^2-2x+5} - \int \frac{x-5}{\sqrt{x^2-2x+5}}\,dx \right)$$

$$= \frac{1}{2}(x-1)\sqrt{x^2-2x+5} + 2\ln\left| x-1 + \sqrt{x^2-2x+5}\right|.$$

the last equality follows from the preceding example.   ■

Let us now consider the integral

(1) $$\int x^p(a+x)^q\,dx$$

(which is sometimes called the integral of a *binomial differential*). Already Newton knew that whenever one of the three numbers p, q, and $p+q$ is an integer this integral can be reduced, by suitable substitution, to an integral of a rational function. We owe to the Russian mathematician Pafnuti Lvovich Chebyshev (1821–1894) proof that these are the only cases in which such a reduction is possible. Suitable substitutions suggest themselves, so we shall consider only a few examples.

6.5.2 Examples

1.
$$I = \int \frac{\sqrt[3]{1 + \sqrt[4]{x}}}{\sqrt{x}}\,dx.$$

This integral has the form $\int x^m(a+x^n)^q\,dx$ which, as one can easily prove, can always be reduced to the form (1). In our case we substitute

$t = \sqrt[3]{1 + \sqrt[4]{x}}$ and get $x = (t^3 - 1)^4$, $x' = 12t^2(t^3 - 1)^3$, and thus

$$I = 12 \int (t^6 - t^3)\, dt = \frac{3}{7}t^4(4t^3 - 7) = \frac{3}{7}\left(1 + \sqrt[4]{x}\right)^{4/3}\left(4\sqrt[4]{x} - 3\right).$$

2.

$$I = \int \frac{dx}{\sqrt[4]{1 + x^4}}.$$

We substitute $x = (t^4 - 1)^{-1/4}$. Hence

$$x' = -t^3(t^4 - 1)^{-5/4}, \qquad \sqrt[4]{1 + x^4} = tx = t(t^4 - 1)^{-1/4}$$

and

$$I = -\int \frac{t^2\, dt}{t^4 - 1} = \frac{1}{4}\int \left(\frac{1}{t+1} - \frac{1}{t-1}\right)dt - \frac{1}{2}\int \frac{dt}{t^2 + 1}$$

$$= \frac{1}{4}\ln\left|\frac{t+1}{t-1}\right| - \frac{1}{2}\arctan t.$$

The integral is obtained by substituting $t = \sqrt[4]{1 + x^4}/x$.

3.

$$I = \int \frac{dx}{x\,\sqrt[3]{1 + x^5}}.$$

Substitution: $x = (t^3 - 1)^{1/5}$, $x' = \frac{3}{5}t^2(t^3 - 1)^{-4/5}$.

$$I = \frac{3}{5}\int \frac{t\, dt}{t^3 - 1} = \frac{1}{5}\int \left(\frac{1}{t-1} - \frac{t-1}{t^2 + t + 1}\right)dt$$

$$= \frac{1}{10}\ln\frac{(t-1)^2}{t^2 + t + 1} + \frac{\sqrt{3}}{5}\arctan\frac{2t+1}{\sqrt{3}}.$$

$\blacksquare$

Exercise 6.5

1. Integrate the following functions:

(a) $\displaystyle\int \frac{x^6}{\sqrt{1 - x^2}}\, dx$; (b) $\displaystyle\int \frac{dx}{(x + 1)^3 \sqrt{x^2 + 2x}}$;

(c) $\displaystyle\int \frac{dx}{x^4 \sqrt{1 + x^2}}$; (d) $\displaystyle\int \frac{dx}{x^2(2 + x^3)^{5/3}}$;

(e) $\displaystyle\int x^{-3/2}(1 + x^{3/4})^{-1/3}\, dx$.

6.6 INTEGRATION OF TRIGONOMETRIC FUNCTIONS

We shall first solve a few particular cases.

6.6.1 Example   Applying the formula $\int f(x) f'(x)\, dx = \tfrac{1}{2}(f(x))^2$ we obtain

$$I = \int \sin x \cos x\, dx = \frac{1}{2}\sin^2 x.$$

The same integral can be solved with the use of the trigonometric identity

$$\sin x \cos x = \frac{1}{2}\sin 2x.$$

$$I = \frac{1}{2}\int \sin 2x\, dx = -\frac{1}{4}\cos 2x.$$

Notice that the obtained answers are not equal; they differ by $1/4$.   ■

The above simple example shows that a variety of methods can be used to solve integrals of trigonometric functions.

6.6.2 Example

$$\int \frac{\sin^5 x}{\cos^4 x}\, dx = -\int \frac{(1 - \cos^2 x)^2 (\cos x)'}{\cos^4 x}\, dx$$

$$= -\int \frac{(\cos x)'}{\cos^4 x}\, dx + 2\int \frac{(\cos x)'}{\cos^2 x}\, dx + \int \sin x\, dx$$

$$= \frac{1}{3\cos^3 x} - \frac{2}{\cos x} - \cos x. \qquad ■$$

6.6.3 Example

$$I = \int \sin^2 x \, \cos^4 x\, dx.$$

We substitute $t = \tan x$ and get the integral of a rational function

$$\int \frac{t^2\, dt}{(1 + t^2)^4}.$$

The same integral can be evaluated in another way. We have

$$\sin^2 x \cos^4 x = \frac{1}{8}\sin^2 2x(\cos 2x + 1) = \frac{1}{8}\sin^2 2x \cos 2x + \frac{1}{16}(1 - \cos 4x).$$

Hence

$$I = \frac{1}{48}\sin^3 2x + \frac{1}{16}x - \frac{1}{64}\sin 4x. \qquad ■$$

6.6.4 Example

$$\int \sin 7x \cos 4x \, dx = \int \frac{1}{2}(\sin 11x + \sin 3x) \, dx = -\frac{1}{22}\cos 11x - \frac{1}{6}\cos 3x.$$

We have used here the following general formula

$$\sin mx \cos nx = \frac{1}{2}\Big(\sin(m + n)x + \sin(m - n)x\Big).$$

The following formulas may be useful in similar integrals.

$$\cos mx \cos nx = \frac{1}{2}\Big(\cos(m + n)x + \cos(m - n)x\Big),$$

$$\sin mx \sin nx = \frac{1}{2}\Big(\cos(m - n)x - \cos(m + n)x\Big).$$

6.6.5 Example

$$I = \int \frac{\sin^2 x \cos x}{\sin x + \cos x} \, dx.$$

Substituting $t = \tan x$ we get

$$I = \int \frac{t^2 \, dt}{(1 + t)(1 + t^2)^2} = \int \Big(\frac{1}{4}\frac{1}{t + 1} - \frac{1}{4}\frac{t - 1}{t^2 + 1} + \frac{1}{2}\frac{t - 1}{(t^2 + 1)^2}\Big) dt$$

$$= \frac{1}{4}\ln\frac{1 + t}{\sqrt{1 + t^2}} - \frac{1}{4}\frac{1 + t}{1 + t^2} = \frac{1}{4}\ln|\cos x + \sin x| - \frac{1}{4}\cos x(\sin x + \cos x).$$

The above example suggests a general method of solving integrals of the form $\int R(\sin x, \cos x)dx$, where R denotes a rational function of two variables. The substitution that works for every R is $t = \tan\frac{1}{2}x \; (-\pi < x < \pi)$, because then

$$\sin x = \frac{2\tan\frac{1}{2}x}{1 + (\tan\frac{1}{2}x)^2} = \frac{2t}{1 + t^2}$$

and

$$\cos x = \frac{1 - (\tan\frac{1}{2}x)^2}{1 + (\tan\frac{1}{2}x)^2} = \frac{1 - t^2}{1 + t^2}.$$

This leads to the integral of the form

$$\int R(\sin x, \cos x) \, dx = \int R\Big(\frac{2t}{1 + t^2}, \frac{1 - t^2}{1 + t^2}\Big)\frac{2 \, dt}{1 + t^2},$$

which is the integral of a rational function, and hence always solvable.

However, although the above method always applies, it often leads to unduly complicated calculations, so before applying it we always should look for another, more suitable, method. This leaves an open field for creativity and individuality. It seems that in the problem of integrating trigonometric functions knowing trigonometry is at least as important as knowing calculus.

Exercise 6.6 ────────────────────────────────────

1. Integrate the following functions:

(a) $\displaystyle\int \frac{\cos^5 x}{\sin^3 x}\,dx;$

(b) $\displaystyle\int \frac{dx}{\sin^2 x \cos^4 x};$

(c) $\displaystyle\int \frac{dx}{\sqrt{\tan x}};$

(d) $\displaystyle\int \sin x \sin 2x \sin 3x\,dx;$

(e) $\displaystyle\int \frac{dx}{\sqrt{\sin x \cos^3 x}};$

(f) $\displaystyle\int \cos\frac{x}{2} \cos\frac{x}{3}\,dx;$

(g) $\displaystyle\int \frac{dx}{\sin x + \cos x};$

(h) $\displaystyle\int \frac{dx}{\cos x + 2\sin x + 3}.$

CHAPTER 7

THE LEBESGUE INTEGRAL

7.1 INTRODUCTION

Measuring the area of geometrical figures on the plane in elementary geometry often consists of dividing them into a finite number of smaller parts like rectangles or triangles, whose areas are easy to calculate, and then summing up the areas of all the parts, Fig. 7.1. This method, however, fails when the figure is bounded by a curve like a circle. In such a case a figure might be decomposed into parts such that each of them is contained between a horizontal line and the arc of a curve. For instance, in Fig. 7.2 there are six such parts. This procedure reduces the evaluation of the area of an arbitrary figure to the evaluation of the area between an arc representing a function and a horizontal line segment.

That evaluation may be undertaken using various methods that lead to the concept of the integral. For continuous functions the integral was already considered by Isaac Barrow (1630–1677), the teacher of Newton. The precise definition is due to Cauchy. Nowadays Georg Friedrich Bernhard Riemann (1826–1866) and Henri-Léon Lebesgue (1875–1941) are best known in the theory of integration. Riemann extended Cauchy's definition so as to include some discontinuous functions. The Riemann integral, however, is unsatisfactory in certain respects. Lebesgue made an essential change in the approach to integration that allows us to deal conveniently with infinite sequences and series of functions.

—Then what is the reason for introducing the Riemann integral?

—Because its definition is easy and requires little preparation. On the contrary, Lebesgue's is usually based on a difficult theory of measure.

—And what will be the approach in this book?

—We shall introduce an integral that is completely equivalent to the Lebesgue integral. But instead of measure theory we shall base it on the theory of absolutely convergent series. The definition is actually simpler than that of the Riemann integral and as general as Lebesgue's. It produces the Lebesgue integral, but is introduced in an elementary way.

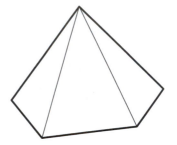

Fig. 7.1 The area can be
calculated by dividing into
triangles.

7.2 INTEGRABLE FUNCTIONS

By a *brick* we shall mean a bounded half-closed interval $[a, b)$. Since bricks are half-closed, when placed one after another they form a single brick . A function that is equal to 1 at the points of a brick J and 0 elsewhere will be called a *brick function* and J its *support*. By the *integral $\int f$ of a brick function f* we mean the length of its support; thus, if the support is $[a, b)$, then $\int f = b - a$.

—The symbol $\int f$ was used before to denote the antiderivative of f. Now $\int f$ is a number. Is this correct?

—No, it is certainly an inconsistency to use the same symbol for different objects, but we shall follow this long tradition, which originates from a connection between the integral and antiderivative. It does not lead to any misunderstanding here since it will always be clear from the context which one we have in mind. In this chapter $\int f$ will always denote the integral of f as a number.

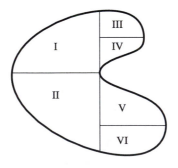

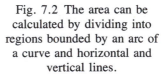

Fig. 7.2 The area can be
calculated by dividing into
regions bounded by an arc of
a curve and horizontal and
vertical lines.

7.2.1 Definition   Given any function f we shall write

(1) $$f \simeq \lambda_1 f_1 + \lambda_2 f_2 + \cdots \quad \text{or} \quad f \simeq \sum_{k=1}^{\infty} \lambda_k f_k$$

where $f_1, f_2, \ldots$ are brick functions and $\lambda_1, \lambda_2, \ldots$ are real numbers, if the following conditions are satisfied:

Ⅰ $\displaystyle\sum_{k=1}^{\infty} |\lambda_k| \int f_k < \infty$, and

Ⅱ $f(x) = \displaystyle\sum_{k=1}^{\infty} \lambda_k f_k(x)$ at all points x at which the series converges absolutely.

If $f \simeq \lambda_1 f_1 + \lambda_2 f_2 + \cdots$, then we say that f *is expandable in a series of brick functions*.

We cannot use the symbol $=$ in place of $\simeq$ because the series need not converge at every point. Note that Ⅰ means simply that the series $\left\{ \sum_{k=1}^{n} |\lambda_k| \int f_k \right\}$ converges. If $\alpha_1, \alpha_2, \ldots$ are positive numbers, then instead of writing "the series $\left\{ \sum_{k=1}^{n} \alpha_k \right\}$ converges" we often use the intuitive expression "$\alpha_1 + \alpha_2 + \cdots < \infty$."

7.2.2 Definition   A function expandable into a series of brick functions is called *integrable*. By the *integral* $\int f$ of an integrable function f satisfying (1) we mean the sum

$$\int f = \sum_{k=1}^{\infty} \lambda_k \int f_k.$$

—This definition is remarkably short and easy to remember. But a function can be expanded in various series of brick functions, so how do we know whether the value of the integral will be always the same?

—At first we do not know. The proof that it is so is not short, not easy to remember, and will not be concluded prior to Section 7.5. To make the matter easier we shall start with a few heuristic remarks. The core of the reasoning is the fact that the integral of a non-negative function is a non-negative number. Before we prove it we need to establish some properties of the so-called step functions. They appear naturally as partial sums of sequences of brick functions.

Exercises 7.2

1. Prove that if $f \simeq \lambda_1 f_1 + \lambda_2 f_2 + \cdots$ and $\alpha \in \mathbb{R}$, then $\alpha f \simeq \alpha \lambda_1 f_1 + \alpha \lambda_2 f_2 + \cdots$.

2. Prove that if $f \simeq \lambda_1 f_1 + \lambda_2 f_2 + \cdots$ and $g \simeq \gamma_1 g_1 + \gamma_2 g_2 + \cdots$ then $f + g \simeq \lambda_1 f_1 + \gamma_1 g_1 + \lambda_2 f_2 + \gamma_2 g_2 + \cdots$.

7.3 EXAMPLES OF EXPANSIONS OF INTEGRABLE FUNCTIONS

We want to show what an expansion into a series of brick functions looks like in a few particular cases. To make the exposition clearer we shall use the following notation. A brick function whose support is $[\alpha, \beta)$ will be denoted by $\chi_{[\alpha,\beta)}$. In general, if S is a subset of $\mathbb{R}$, then by χ_S we denote the *characteristic function* of S, namely,

$$\chi_S(x) = \begin{cases} 1 & \text{for } x \in S \\ 0 & \text{for } x \notin S \end{cases}$$

7.3.1 Example   The function

$$\chi_{\{\alpha\}}(x) = \begin{cases} 1 & \text{for } x = \alpha \\ 0 & \text{for } x \neq \alpha \end{cases}$$

(Fig. 7.3) is an example of an integrable function. It is easy to construct an expansion that converges to $\chi_{\{\alpha\}}$ at every point. Indeed, let $\{\beta_n\}$ be a sequence of numbers decreasing to α. Then

$$\chi_{\{\alpha\}} \simeq \chi_{[\alpha,\beta_1)} - \chi_{[\beta_2,\beta_1)} - \chi_{[\beta_3,\beta_2)} - \cdots.$$

According to the definition, we have

$$\int \chi_{\{\alpha\}} = (\beta_1 - \alpha) - (\beta_1 - \beta_2) - (\beta_2 - \beta_3) - \cdots = 0. \qquad \blacksquare$$

7.3.2 Example   The continuous function whose graph is seen in Fig. 7.4 can be analytically defined as

$$f(x) = \begin{cases} 1 + x & \text{for } -1 \le x < 0 \\ 1 - x & \text{for } 0 \le x < 1 \\ 0 & \text{elsewhere.} \end{cases}$$

This function is integrable. It can be expanded as follows:

$$f \simeq \frac{1}{2}\chi_{[-\frac{1}{2},\frac{1}{2})} + \frac{1}{4}\chi_{[-\frac{3}{4},-\frac{2}{4})} + \frac{1}{4}\chi_{[-\frac{1}{4},\frac{1}{4})} + \frac{1}{4}\chi_{[\frac{2}{4},\frac{3}{4})} + \frac{1}{8}\chi_{[-\frac{7}{8},-\frac{6}{8})} + \frac{1}{8}\chi_{[-\frac{5}{8},-\frac{4}{8})}$$

$$+ \frac{1}{8}\chi_{[-\frac{3}{8},-\frac{2}{8})} + \frac{1}{8}\chi_{[-\frac{1}{8},\frac{1}{8})} + \frac{1}{8}\chi_{[\frac{2}{8},\frac{3}{8})} + \frac{1}{8}\chi_{[\frac{4}{8},\frac{5}{8})} + \frac{1}{8}\chi_{[\frac{6}{8},\frac{7}{8})} + \cdots.$$

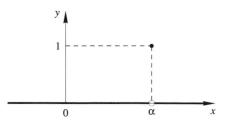

Fig. 7.3 A discontinuous integrable
function.

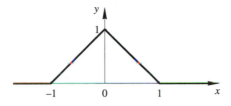

Fig. 7.4 A continuous integrable
function.

The partial sums consisting of one, four, and eleven terms are represented in Fig. 7.5. According to the definition of the integral we have

$$\int f = \frac{1}{2} + \frac{1}{16} + \frac{1}{8} + \frac{1}{16} + \frac{1}{64} + \frac{1}{64} + \frac{1}{64} + \frac{1}{32} + \frac{1}{64} + \frac{1}{64} + \frac{1}{64} + \cdots$$

$$= \frac{1}{2} + \frac{1}{4} + \frac{1}{8} + \cdots = 1.$$

A similar construction can be made for any integrable continuous function (see Theorem 7.14.3). Of course, the construction is, in general, much more complicated. ∎

7.3.3 Example   Let $\left\{ \sum_{k=1}^{n} \lambda_k \right\}$ be a convergent series of positive numbers. The function

$$f(x) = \begin{cases} \lambda_n & \text{for } n \le x < n+1 \\ 0 & \text{elsewhere} \end{cases}$$

is an integrable function. The expansion suggests itself

$$f \simeq \lambda_1 \chi_{[1,2)} + \lambda_2 \chi_{[2,3)} + \cdots.$$

Hence we have $\int f = \lambda_1 + \lambda_2 + \cdots.$ ∎

Exercises 7.3

1.   Let

$$f(x) = \begin{cases} 1 & \text{for } x = 0 \\ 0 & \text{for } x \ne 0 \end{cases}$$

Verify whether the following are valid expansions of f.

(a) $f \simeq \chi_{[0,\frac{1}{2})} - \chi_{[0,\frac{1}{2})} + \chi_{[0,\frac{1}{4})} - \chi_{[0,\frac{1}{4})} + \cdots + \chi_{[0,2^{-n})} - \chi_{[0,2^{-n})} + \cdots.$

(b) $f \simeq \chi_{[0,\frac{1}{2})} - \chi_{[0,\frac{1}{2})} + 2\chi_{[0,\frac{1}{4})} - 2\chi_{[0,\frac{1}{4})} + \cdots + n\chi_{[0,2^{-n})} - n\chi_{[0,2^{-n})} + \cdots.$

(c) $f \simeq \chi_{[0,1)} - \chi_{[0,1)} + \chi_{[0,\frac{1}{2})} - \chi_{[0,\frac{1}{2})} + \cdots + \chi_{[0,\frac{1}{n})} - \chi_{[0,\frac{1}{n})} + \cdots.$

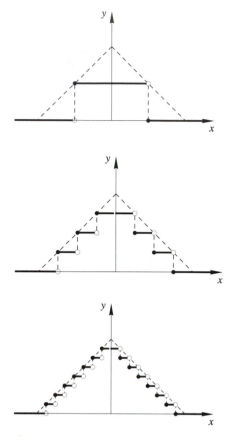

Fig. 7.5 Partial sums of an expansion of the function in Fig. 7.4.

(d) $\quad f \simeq \chi_{[0,1)} - \chi_{[0,1)} + \dfrac{1}{2}\chi_{[0,\frac{1}{2})} - \dfrac{1}{2}\chi_{[0,\frac{1}{2})} + \cdots + \dfrac{1}{n}\chi_{[0,\frac{1}{n})} - \dfrac{1}{n}\chi_{[0,\frac{1}{n})} + \cdots.$

(e) $\quad f \simeq \dfrac{1}{2}\chi_{[0,1)} - \dfrac{1}{2}\chi_{[0,1)} + \dfrac{1}{4}\chi_{[0,\frac{1}{2})} - \dfrac{1}{4}\chi_{[0,\frac{1}{2})} + \cdots + \dfrac{1}{2^n}\chi_{[0,\frac{1}{n})} - \dfrac{1}{2^n}\chi_{[0,\frac{1}{n})} + \cdots.$

2. Is the expansion

$$f \simeq \chi_{[0,1)} - \chi_{[0,\frac{1}{\sqrt{2}})} + \chi_{[0,\frac{1}{\sqrt{3}})} - \cdots + (-1)^{n+1}\chi_{[0,\frac{1}{\sqrt{n}})} + \cdots.$$

valid?

3. Find brick-function expansions of the following functions and calculate the integrals.

   (a) $\chi_{[a,b]}$;

   (b) $\chi_{(a,b)}$;

(c) $\chi_{(a,b]}$;

(d) $f(x) = \begin{cases} 0 & \text{if } x < 0 \\ x & \text{if } 0 \le x \le 1 \\ 0 & \text{if } x > 1. \end{cases}$

4.   Let f be an integrable function and let $\alpha, \beta \in \mathbb{R}$, $\alpha \ne 0$. Define $g(x) = f(\alpha x + \beta)$. Prove that g is an integrable function and find $\int g$.

7.4 STEP FUNCTIONS

7.4.1 Definition   A linear combination of a finite number of brick functions is called a *step function*.

According to the above definition, a function f is a step function if there exist numbers $\lambda_1, \ldots, \lambda_n$ and brick functions $f_1, \ldots, f_n$ such that

(1) $f(x) = \lambda_1 f_1(x) + \cdots + \lambda_n f_n(x)$ for every $x \in \mathbb{R}$.

An example of a step function is shown in Fig. 7.6. Evidently, every step function can be represented in the form (1) in infinitely many ways. On the other hand, it is quite clear that there is exactly one representation in which the supports of $f_1, \ldots, f_n$ are disjoint and the minimal number of brick functions is used. To obtain such a representation, denote by $a_0, \ldots, a_m$ all points of discontinuity of f arranged into an increasing sequence, that is, $a_0 < \cdots < a_m$. Then, for $k = 1, \ldots, m$, define g_k to be the characteristic function of $[a_{k-1}, a_k)$, and by γ_k the value of f in that interval, that is, $\gamma_k = f(a_{k-1})$. Then $f = \gamma_1 g_1 + \cdots + \gamma_m g_m$ is the desired representation. Such a representation will be called the *basic representation* of f.

From the definition it follows that $f = 0$ is a step function. However, in this case a basic representation does not make much sense. For convenience, $f = 0$ will be called the basic representation of this function.

The set of all $x \in \mathbb{R}$ for which $f(x) \ne 0$ is called the support of f and denoted supp f.

It is easily seen that a linear combination of a finite number of step functions is a step function. Also the absolute value of a step function is a step function.

Let f and g be step functions. Since

$$\min(f, g) = \frac{1}{2}(f + g - |f - g|) \quad \text{and} \quad \max(f, g) = \frac{1}{2}(f + g + |f - g|),$$

the minimum and maximum of two step functions are step functions, too.

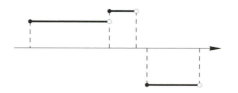

Fig. 7.6 A step function.

7.4.2 Definition   By the *integral ∫f of a step function f* with representation (1) we mean

$$\int f = \lambda_1 \int f_1 + \cdots + \lambda_n \int f_n.$$

It is rather obvious that the defined integral is independent of a particular representation (1), but a formal proof would be lengthy. Using this independence we can easily prove the following simple but important properties of step functions:

$$\int (f + g) = \int f + \int g.$$

$$\int \lambda f = \lambda \int f, \lambda \in \mathbb{R}.$$

$$f \le g \text{ implies } \int f \le \int g.$$

$$\left| \int f \right| \le \int |f|.$$

If $|f| \le M$ and the support of f is contained in $[a, b]$, then $\left| \int f \right| \le (b - a)M$.

We close this section with three auxiliary lemmas. The first lemma is used only to prove the second one. The second one is used to prove the third, but it is also interesting in itself. The result of the third lemma is crucial in the proof of the uniqueness of the integral defined in Section 7.2.

7.4.3 Lemma   *Given any non-negative step function f and a number $\varepsilon > 0$ there exists a non-negative step function g and a number $\eta > 0$ such that*

(2) $g(x) \le f(y)$ *for all x and y such that $|x - y| < \eta$,*

and

(3) $\int (f - g) < \varepsilon.$

The method of construction of a function g with the desired properties is shown in Fig. 7.7. The continuous line represents the graph of f, and the broken line the graph of g. A formal proof can be produced as follows:

Proof.   Let $a_0, \ldots, a_n$ be the points of discontinuity of f and let δ be a positive number small enough to make the intervals $[a_0 - \delta, a_0 + \delta], \ldots, [a_n - \delta, a_n + \delta]$ disjoint. Then f is constant on each of intervals $(a_k - \delta, a_k)$ and $(a_k, a_k + \delta)$. Set $c_k = f(a_k - \frac{1}{2}\delta)$ and $d_k = f(a_k + \frac{1}{2}\delta)$, $k = 0, \ldots, n$. Now we can define g as

$$g(x) = \begin{cases} \min(c_k, d_k) & \text{if } x \in (a_k - \eta, a_k + \eta) \text{ for some } k \in \{0, 1, \ldots, n\} \\ f(x) & \text{otherwise} \end{cases}$$

where η is a positive number chosen so that (2) and (3) are satisfied. To satisfy (2)

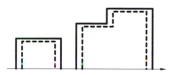

Fig. 7.7 Illustration for the
proof of Lemma 7.4.3.

it suffices to take $\eta < \delta$. Since

$$\int (f - g) = \sum_{k=0}^{n} |c_k - d_k| \eta,$$

if

$$\eta < \min \left\{ \frac{\varepsilon}{\sum\limits_{k=0}^{n} |c_k - d_k|}, \delta \right\},$$

then both conditions (2) and (3) are satisfied.

The statement in the following lemma may sound obvious. However, careful analysis of the described property reveals that it is not obvious at all. The proof of this lemma is probably the most difficult one in this book.

7.4.4 Lemma   *If $\{f_n\}$ is a nonincreasing sequence of step functions that converges to 0 at every point, then $\int f_n \to 0$ as $n \to \infty$.*

Proof.   First notice that, since the step functions f_n are nonnegative and the sequence $\{f_n\}$ is nonincreasing, the sequence $\{\int f_n\}$ is non-increasing and bounded from below by 0. Consequently, the sequence $\{\int f_n\}$ converges. We have to show that the limit is 0.

By Lemma 7.4.3, for every $\varepsilon > 0$ and for every $n \in \mathbb{N}$ there exist a step function g_n and a number $\eta_n > 0$ such that

(4) $\qquad\qquad g_n(x) \le f_n(y) \quad$ for all x and y such that $|x - y| < \eta_n$,

and

(5) $$\int f_n < \int g_n + \frac{\varepsilon}{2^n}.$$

Obviously, the numbers η_n can be chosen to form a decreasing sequence.
Next define

$$h_1 = g_1$$

and,

$$h_n = g_n - (f_1 - g_1) - (f_2 - g_2) - \ldots - (f_{n-1} - g_{n-1})$$

for $n = 2, 3, \ldots$. Then

(6) $\qquad h_n(x) \leq f_n(y) \quad$ for all x and y such that $|x - y| < \eta_n$,

(7) $\qquad \displaystyle\int f_n < \int h_n + \varepsilon,$

(8) $\qquad h_{n+1} \leq h_n,$

for all $n \in \mathbb{N}$. Indeed, by (4), we have $h_n \leq g_n$ and thus

$$h_n(x) \leq g_n(x) \leq f_n(y) \quad \text{whenever } |x - y| < \eta_n.$$

Moreover, by (5), we have

$$\int (f_n - h_n) \leq \int (f_n - g_n) + \int (f_1 - g_1) + \cdots + \int (f_{n-1} - g_{n-1}) < \varepsilon,$$

proving (7). Finally,

$$
\begin{aligned}
h_{n+1} &= g_{n+1} - (f_1 - g_1) - (f_2 - g_2) - \ldots - (f_n - g_n) \\
&= g_n - (f_1 - g_1) - (f_2 - g_2) - \ldots - (f_{n-1} - g_{n-1}) + (g_{n+1} - f_n) \\
&= h_n + g_{n+1} - f_n \\
&\leq h_n + g_{n+1} - f_{n+1} \leq h_n,
\end{aligned}
$$

because $\{f_n\}$ is nonincreasing and $g_{n+1} - f_{n+1} \leq 0$.

We will prove now that for every $\delta > 0$ there exists $n_\delta \in \mathbb{N}$ such that

(9) $\qquad h_n \leq \delta \quad$ for every $n \geq n_\delta.$

Suppose this is not true for some $\delta > 0$. Let, for $n = 1, 2, \ldots, x_n$ be a point where h_n assumes its maximum. Then there exists a subsequence $\{h_{p_n}\}$ of $\{h_n\}$ such that

(10) $\qquad h_{p_n}(x_{p_n}) > \delta$

for all $n \in \mathbb{N}$. Since the supports of all the h_n are contained in some bounded interval $[a, b]$, by the Bolzano–Weierstrass theorem there exists a subsequence $\{x_{q_n}\}$ of $\{x_{p_n}\}$ that converges to some $x_0 \in [a, b]$. Since $f_n(x_0) \to 0$, there exists an index $m \in \mathbb{N}$ such that

(11) $\qquad f_m(x_0) < \delta.$

On the other hand, since $x_{q_n} \to x_0$, there exists an index $q_k \geq m$ such that

$$|x_{q_k} - x_0| < \eta_m$$

and thus

(12) $\qquad h_m(x_{q_k}) \leq f_m(x_0).$

Since $\{h_n\}$ is a nonincreasing sequence, (12) and $q_k \geq m$ imply that

(13) $\qquad h_{q_k}(x_{q_k}) \leq h_m(x_{q_k}).$

Combining (11), (12), and (13) we obtain

$$h_{q_k}(x_{q_k}) \leq \delta.$$

But this contradicts (10), proving that (9) holds for every δ.

Since $\int h_{n+1} \leq \int h_n$, by (8), and $-\varepsilon \leq \int f_n - \varepsilon \leq \int h_n$, by (7), the limit $\lim_{n \to \infty} \int h_n$ exists. Moreover, since $\mathrm{supp}\, h_n \subset [a, b]$ for all $n \in \mathbb{N}$, we have $\int h_n \leq \delta(b - a)$ for $n \geq n_\delta$, by (9). Finally, since δ can be an arbitrary small positive number, we conclude that

$$(14) \qquad \lim_{n \to \infty} \int h_n \leq 0.$$

Now, by (7) and (14), we have

$$0 \leq \lim_{n \to \infty} \int f_n \leq \lim_{n \to \infty} \int h_n + \varepsilon \leq \varepsilon$$

Since ε is an arbitrary positive number, we obtain

$$\lim_{n \to \infty} \int f_n = 0,$$

completing the proof.

7.4.5 Lemma   *If $\{f_n\}$ and $\{g_n\}$ are nondecreasing sequences of step functions and*

$$\lim_{n \to \infty} f_n(x) \leq \lim_{n \to \infty} g_n(x)$$

for every $x \in \mathbb{R}$, then

$$\lim_{n \to \infty} \int f_n(x) \leq \lim_{n \to \infty} \int g_n(x).$$

(In both inequalities we allow $a \leq \infty$ ($a \in \mathbb{R}$) and $\infty \leq \infty$).

Proof.   Let $m \in \mathbb{N}$ be fixed for now. Set

$$(15) \qquad h_n = g_n - f_m.$$

We decompose h_n into its positive and negative parts

$$(16) \qquad h_n = h_n^+ - h_n^-,$$

where $h_n^+ = \max(h_n, 0)$ and $h_n^- = \max(-h_n, 0)$. The sequence $\{h_n^-\}$ is nonincreasing and $\lim_{n \to \infty} h_n^- = 0$. Thus by Lemma 7.4.4 we have

$$(17) \qquad \lim_{n \to \infty} \int h_n^- = 0.$$

From (16) it follows that $\int h_n = \int h_n^+ - \int h_n^-$, and, by (17), $\lim_{n \to \infty} \int h_n = \lim_{n \to \infty} \int h_n^+ \geq 0$. But (15) implies that $\int h_n = \int (g_n - f_m)$, and thus $\lim_{n \to \infty} \int g_n - \int f_m \geq 0$; that is, $\int f_m \leq \lim_{n \to \infty} \int g_n$. Now, by letting $m \to \infty$, we obtain

$$\lim_{m \to \infty} \int f_m \leq \lim_{n \to \infty} \int g_n,$$

which is the desired inequality.

Exercises 7.4

1. Find basic representations for the following step functions:
   (a) $f = \chi_{[-2,2)} - 2\chi_{[-1,1)}$,
   (b) $g = \chi_{[0,2)} - 3\chi_{[1,3)}$,
   (c) $h = \chi_{[0,5)} + 2\chi_{[1,4)} - \chi_{[2,3)}$.

2. True or false? If $f = \lambda_1 f_1 + \cdots + \lambda_n f_n$ is the basic representation of f, then $|f| = \lambda_1|f_1| + \cdots + \lambda_n|f_n|$ is the basic representation of $|f|$.

3. Prove the following properties of the step functions:
   (a) $\int (f + g) = \int f + \int g$,
   (b) $\int \lambda f = \lambda \int f, \lambda \in \mathbb{R}$,
   (c) $f \le g$ implies $\int f \le \int g$,
   (d) $|\int f| \le \int |f|$.

4. Let f be a step function. Prove that if $|f| \le M$ and the support of f is contained in $[a, b)$, then $|\int f| \le (b - a)M$.

5. Let $f = \chi_{[0,2)} + \chi_{[1,2)} + 3\chi_{[3,4)}$. Find a step function g and a positive number η such that $g(x) \le f(y)$ for all x and y satisfying $|x - y| < \eta$, and $\int (f - g) < \frac{1}{7}$.

6. Prove that if $\{f_n\}$ and $\{g_n\}$ are nonincreasing sequences of step functions and $\lim_{n\to\infty} f_n(x) \le \lim_{n\to\infty} g_n(x)$ for every $x \in \mathbb{R}$, then $\lim_{n\to\infty} \int f_n(x) \le \lim_{n\to\infty} \int g_n(x)$.

7. Prove that if $\{f_n\}$ and $\{g_n\}$ are nondecreasing sequences of step functions and $\lim_{n\to\infty} f_n(x) = \lim_{n\to\infty} g_n(x)$ for every $x \in \mathbb{R}$, then $\lim_{n\to\infty} \int f_n(x) = \lim_{n\to\infty} \int g_n(x)$.

8. True or false? If $\{f_n\}$ is a sequence of step functions that converges to 0 at every point, then $\lim_{n\to\infty} \int f_n = 0$.

9. If f is integrable and vanishes outside of a bounded interval J, then there are step functions $f_1, f_2, \ldots$, vanishing outside of J such that $f \simeq f_1 + f_2 + \cdots$.

10. Show that for any functions f and g;
   (a) supp $(f + g) \subseteq$ supp $f \cup$ supp g;
   (b) supp $fg =$ supp $f \cap$ supp g;
   (c) supp $|f| =$ supp f;
   (d) supp $\lambda f =$ supp f, $\lambda \in \mathbb{R}$, $\lambda \ne 0$;
   (e) If $|f| \le |g|$ then supp $f \subseteq$ supp g.

7.5 BASIC PROPERTIES OF THE INTEGRAL

First we extend the use of the symbol $\simeq$. Let f be an arbitrary function and let $\{f_n\}$ be a sequence of step functions. We write

(1)
$$f \simeq f_1 + f_2 + \cdots \quad \text{or} \quad f \simeq \sum_{k=1}^{\infty} f_k$$

if

Ⅰ $\sum_{k=1}^{\infty} \int |f_k| < \infty$, and

Ⅱ $f(x) = \sum_{k=1}^{\infty} f_k(x)$ at all points x at which the series converges absolutely.

If $f \simeq f_1 + f_2 + \cdots$, then we say that f is expandable into a series of step functions. We have the following important theorem:

7.5.1 Theorem   *A function is integrable if and only if it can be expanded into a series of step functions. Moreover, if $f \simeq f_1 + f_2 + \cdots$, where the f_n are step functions, then $\int f = \int f_1 + \int f_2 + \cdots$.*

Proof.   If $f \simeq \lambda_1 f_1 + \lambda_2 f_2 + \cdots$, where f_n are brick functions, then f is expandable into the series of step functions $\lambda_n f_n$. Suppose now that $f \simeq f_1 + f_2 + \cdots$ is an expansion of f into a series of step functions. Let, for $n = 1, 2, \ldots,$

$$f_n = \lambda_{n1} f_{n1} + \cdots + \lambda_{np_n} f_{np_n}$$

be the basic representation of f_n. Then

$$f \simeq \lambda_{11} f_{11} + \cdots + \lambda_{1p_1} f_{1p_1} + \cdots + \lambda_{n1} f_{n1} + \cdots + \lambda_{np_n} f_{np_n} + \cdots.$$

In fact, since

$$\int |f_n| = |\lambda_{n1}| \int f_{n1} + \cdots + |\lambda_{np_n}| \int f_{np_n},$$

condition Ⅰ is satisfied. Moreover, if, for some $x \in \mathbb{R}$,

$$|\lambda_{11}| f_{11}(x) + \cdots + |\lambda_{1p_1}| f_{1p_1}(x) + \cdots + |\lambda_{n1}| f_{n1}(x) + \cdots + |\lambda_{np_n}| f_{np_n}(x) + \cdots < \infty,$$

then

$$|f_1(x)| + |f_2(x)| + \cdots < \infty,$$

and thus

$$f(x) = \lambda_{11} f_{11}(x) + \cdots + \lambda_{1p_1} f_{1p_1}(x) + \cdots + \lambda_{n1} f_{n1}(x) + \cdots + \lambda_{np_n} f_{np_n}(x) + \cdots.$$

which proves that condition Ⅱ is also satisfied. Therefore, f is integrable. Moreover,

$$\int f = \lambda_{11} \int f_{11} + \cdots + \lambda_{1p_1} \int f_{1p_1} + \cdots + \lambda_{n1} \int f_{n1} + \cdots \lambda_{np_n} \int f_{np_n} + \cdots$$

$$= \int f_1 + \int f_2 + \cdots,$$

which completes the proof.

Note that the above lemma shows that in the definition of the Lebesgue integral step functions can be used instead of brick functions. The reasons for choosing brick functions in the original definition are that brick functions are simpler, yet sufficient, and that expansion into a series of brick functions corresponds better to the geometrical intuition behind the definition. On the other hand, in many arguments it is much more convenient to use expansions into series of step functions. In the remainder of this section and in Section 7.6 all theorems are to be understood as being formulated for series of step functions.

7.5.2 Theorem   *If $f \simeq f_1 + f_2 + \cdots$ and $f \geq 0$, then $\int f_1 + \int f_2 + \cdots \geq 0$.*

Proof.   Let $\varepsilon > 0$. By ▯ there exists an $n_0 \in \mathbb{N}$ such that

$$\sum_{k=n_o+1}^{\infty} \int |f_k| < \varepsilon.$$

For $n = 1, 2, \ldots$, define

$$g_n = f_1 + \cdots + f_{n_o} + |f_{n_o+1}| + \cdots + |f_{n_o+n}|$$

and

$$h_n = \max(g_n, 0).$$

Clearly, the g_n and h_n are step functions and the sequences $\{g_n\}$ and $\{h_n\}$ are nondecreasing. Moreover, since $f \geq 0$, ▯▯ implies that the limit $\lim_{n\to\infty} g_n(x)$ is either a non-negative number or ∞. Therefore, $\lim_{n\to\infty} g_n(x) = \lim_{n\to\infty} h_n(x)$ for every $x \in \mathbb{R}$. Thus, by Lemma 7.4.5, we have

$$\lim_{n\to\infty} \int g_n(x) \geq \lim_{n\to\infty} \int h_n(x) \geq 0.$$

Consequently

$$\int f_1 + \cdots + \int f_{n_o} + \int |f_{n_o+1}| + \int |f_{n_o+2}| + \cdots \geq 0,$$

Since

$$\int f_1 + \cdots + \int f_{n_o} - \int |f_{n_o+1}| - \int |f_{n_o+2}| + \cdots \leq \sum_{k=1}^{\infty} \int f_k,$$

$$\sum_{k=1}^{\infty} \int f_k \leq \int f_1 + \cdots + \int f_{n_o} + \int |f_{n_o+1}| + \int |f_{n_o+2}| + \cdots,$$

and

$$\sum_{k=n_o+1}^{\infty} \int |f_k| < \varepsilon,$$

we have

$$\sum_{k=1}^{\infty} \int f_k \geq -2\varepsilon.$$

Because ε is an arbitrary positive number, we conclude that $\sum_{k=1}^{\infty} \int f_k \geq 0$.

From the above theorem we can easily obtain the uniqueness of the integral in Definition 7.2.2.

7.5.3 Theorem   *If* $f \simeq f_1 + f_2 + \cdots$ *and* $f \simeq g_1 + g_2 + \cdots$, *then*

$$\int f_1 + \int f_2 + \cdots = \int g_1 + \int g_2 + \cdots.$$

Proof.   Since

$$0 \simeq f_1 - g_1 + f_2 - g_2 + \cdots$$

we have, in view of Theorem 7.5.2,

$$0 \leq \int 0 = \int f_1 - \int g_1 + \int f_2 - \int g_2 + \cdots$$

Thus

$$\int f_1 + \int f_2 + \cdots \geq \int g_1 + \int g_2 + \cdots.$$

Since the roles of both expansions are symmetric, we similarly obtain

$$\int f_1 + \int f_2 + \cdots \leq \int g_1 + \int g_2 + \cdots,$$

which proves the desired equality.

Only now is the use of the symbol $\int f$ justified for both the integral of a step function f and the integral of the same function f defined by an expansion into a series of brick functions. Moreover, the uniqueness of the integral being proved, it follows directly from the definition that

$$\int (\lambda f) = \lambda \int f \quad \text{and} \quad \int (f + g) = \int f + \int g.$$

From Theorem 7.5.2 we immediately derive that

$$f \leq g \quad \text{implies} \quad \int f \leq \int g.$$

—In the definition of integrable functions the only connection between the function f and the series of brick functions is at those points at which the series is absolutely convergent. What if there are no such points?

—This is impossible. It can be proved that in every brick there are points at which the series converges absolutely. In fact, suppose that $f \simeq \lambda_1 f_1 + \lambda_2 f_2 + \cdots$ and that there is a brick $[a, b)$ such that the series $\{\sum_{k=1}^{n} \lambda_k f_k(x)\}$ does not converge absolutely at any point of $[a, b)$. Let f_0 be the characteristic function of $[a, b)$. Then we have $f \simeq f_0 + \lambda_1 f_1 + \lambda_2 f_2 + \cdots$ and hence $\int f = \int f_0 + \lambda_1 \int f_1 + \lambda_2 \int f_2 + \cdots$. Since we know that $\int f = \lambda_1 \int f_1 + \lambda_2 \int f_2 + \cdots$, the uniqueness of the integral implies that $\int f_0 = 0$, which contradicts the definition of f_0.

—Thus the set of those "bad" points cannot be too large.

—We can prove that if f_0 is the characteristic function of the set Z of all points x for which the series $\{\sum_{k=1}^{n} \lambda_k f_k(x)\}$ does not converge absolutely, then $\int f_0 = 0$. In fact, since $f - f_0 \simeq \lambda_1 f_1 + \lambda_2 f_2 + \cdots$, we have $\int (f - f_0) = \int f$. Hence $\int f_0 = \int f - \int (f - f_0) = 0$. We say then that Z is a *null set*. Generally, we say that a set is a null set if its characteristic function is integrable and the integral equals 0.

Exercises 7.5

1. Prove that if $f \simeq f_1 + f_2 + \cdots$ and g is a step function, then
$$f + g \simeq g + f_1 + f_2 + \cdots .$$

2. Prove that if $f \simeq f_1 + f_2 + \cdots$ and $g \simeq g_1 + g_2 + \cdots$ then
$$f \simeq f_1 + g_1 - g_1 + f_2 + g_2 - g_2 + f_3 + g_3 - g_3 + \cdots .$$

7.6 THE MODULUS OF AN INTEGRABLE FUNCTION

The following theorem plays a very important role in the theory of the integral.

7.6.1 Theorem   *The modulus $|f|$ of an integrable function*

(1)
$$f \simeq f_1 + f_2 + \cdots$$

is integrable and

(2)
$$\left| \int f_1 + \int f_2 + \cdots \right| \le \int |f| \le \int |f_1| + \int |f_2| + \cdots$$

Proof.   The equality
$$f(x) = f_1(x) + f_2(x) + \cdots$$

holds at every point at which the series converges absolutely. Let Z denote the set of all those points. Then
$$f(x) = \lim_{n \to \infty} s_n(x) \quad \text{for all } x \in Z,$$

where
$$s_n = f_1 + \cdots + f_n .$$

Consequently
$$|f(x)| = \lim_{n \to \infty} |s_n(x)| \quad \text{for all } x \in Z.$$

Denote $g_1 = |s_1| = |f_1|$, $g_2 = |s_2| - |s_1|$, and generally
$$g_n = |s_n| - |s_{n-1}| \quad \text{for } n \ge 2.$$

Note that all the g_n (as well as s_n) are step functions. Since

(3)
$$|g_n| \leq \big||s_n| - |s_{n-1}|\big| \leq |s_n - s_{n-1}| = |f_n|,$$

we have

(4)
$$\sum_{k=1}^{\infty} \int |g_k| \leq \sum_{k=1}^{\infty} \int |f_k| < \infty.$$

Although

$$|f(x)| = \sum_{k=1}^{\infty} g_k(x) \quad \text{for all } x \in Z,$$

we cannot claim that $|f| \simeq g_1 + g_2 + \cdots$, because the series may converge to extraneous values at some points outside Z. In order to "spoil" the convergence of the series at those points we modify the series by adding and subtracting terms of the series $\left\{ \sum_{k=1}^{n} f_k \right\}$ to get

$$|f| \simeq g_1 + f_1 - f_1 + g_2 + f_2 - f_2 + \cdots.$$

Indeed, condition Ⅰ is still satisfied by (4). Moreover, the above series and the series in (1) converge absolutely at exactly the same points. Therefore, condition Ⅱ is also satisfied, which proves that $|f|$ is integrable, in view of Theorem 7.5.1.

To prove the first inequality in (2) note that $f \leq |f|$ and that $-f \leq |f|$, and hence $\int f \leq \int|f|$ and $-\int f \leq \int|f|$, which yields $|\int f| \leq \int|f|$. Finally,

$$\int |f| = \int g_1 + \int g_2 + \cdots \leq \int |f_1| + \int |f_2| + \cdots,$$

by (3). The proof is thus completed.

Now we shall prove a theorem that will be used in the proof of an important theorem in the next section. The result is also interesting in itself.

7.6.2 Theorem   *For any integrable function f and any positive number ε there exists an expansion into a series of step functions $f \simeq f_1 + f_2 + \cdots$ such that*

$$\int |f_1| + \int |f_2| + \cdots < \int |f| + \varepsilon.$$

Proof.   Let us first take an arbitrary expansion

(5)
$$f \simeq g_1 + g_2 + \cdots.$$

By Ⅰ, there exists a number $n_0 \in \mathbb{N}$ such that

(6)
$$\sum_{k=n_o+1}^{\infty} \int |g_k| < \frac{1}{2}\varepsilon.$$

Define

$$f_1 = g_1 + \cdots + g_{n_o}, \qquad f_2 = g_{n_o+1}, \qquad f_3 = g_{n_o+2},$$

and in general

(7) $$f_n = g_{n_0+n-1} \quad \text{for } n \geq 2.$$

Then clearly

$$f \simeq f_1 + f_2 + \cdots.$$

and hence also

(8) $$f - f_1 \simeq f_2 + f_3 + \cdots$$

Moreover, by (6) and (7), we have

(9) $$\int |f_2| + \int |f_3| + \cdots < \frac{1}{2}\varepsilon$$

The inequality

$$|f_1| \leq |f| + |f - f_1|$$

implies that

$$\int |f_1| \leq \int |f| + \int |f - f_1|.$$

Now, applying Theorem 7.6.1 to the function $f - f_1$ and using (8), we obtain

$$\int |f_1| \leq \int |f| + \int |f_2| + \int |f_3| + \cdots.$$

Hence, by (9),

$$\int |f_1| \leq \int |f| + \frac{1}{2}\varepsilon.$$

By adding this inequality and (9) we get the desired result

$$\int |f_1| + \int |f_2| + \cdots < \int |f| + \varepsilon.$$

Exercises 7.6

1. Prove that if f is integrable, then $\max\{0, f\}$ and $\min\{0, f\}$ are integrable.
2. Show that the relation

$$f \sim g \quad \text{if and only if} \quad \int |f - g| = 0$$

   is an equivalence relation in the space of integrable functions.
3. If f is an integrable function and $f \geq 0$, is it always possible to find nonnegative step functions $f_1, f_2, \ldots$ such that $f \simeq f_1 + f_2 + \cdots$?
4. Show that if f is a continuous integrable function, then there are step functions $f_1, f_2, \ldots$ such that $f \simeq f_1 + f_2 + \cdots$ and $|f| \simeq |f_1| + |f_2| + \cdots$.

7.7 SERIES OF INTEGRABLE FUNCTIONS

In this section we extend the use of the symbol $\simeq$ once again. This will be the final extension.

By an expansion

$$f \simeq f_1 + f_2 + \cdots$$

of a function f in a series of integrable functions $f_1, f_2, \ldots$ we mean a series such that

  I $\displaystyle\sum_{k=1}^{\infty} \int |f_k| < \infty$, and

  II $f(x) = \displaystyle\sum_{k=1}^{\infty} f_k(x)$ at all points x at which the series converges absolutely.

Note that the only difference between the definitions of $\simeq$ given in Sections 7.2 and 7.5 and the above definition is the class of functions allowed in the series. At the beginning only brick functions were allowed, then step functions, and finally any integrable functions.

The following theorem is of fundamental importance in the theory of the integral.

7.7.1 Theorem   *If a function f is expandable into a series of integrable functions*

(1) $$f \simeq f_1 + f_2 + \cdots.$$

then it is integrable and

(2) $$\int f = \int f_1 + \int f_2 + \cdots$$

Proof.   Let $\left\{\sum_{k=1}^{n} \varepsilon_k\right\}$ be a convergent series of positive numbers. By Theorem 7.6.2 there are expansions

(3) $$f_n \simeq f_{n1} + f_{n2} + \cdots \quad (n = 1, 2, \ldots)$$

into series of step functions such that

(4) $$\int |f_{n1}| + \int |f_{n2}| + \cdots < \int |f_n| + \varepsilon_n.$$

Let

(5) $$\left\{\sum_{k=1}^{n} g_k\right\}$$

be a series of step functions arranged from all f_{nk} $(n, k \in \mathbb{N})$. Then, by (4) and Theorem 4.6.6, we have

(6) $$\int |g_1| + \int |g_2| + \cdots < \int |f_1| + \int |f_2| + \cdots + \varepsilon_1 + \varepsilon_2 + \cdots < \infty.$$

Moreover, if series (5) converges absolutely at a point x, then all the series in (3) as well as the series in (1) converge absolutely at x and

$$g_1(x) + g_2(x) + \cdots = f_1(x) + f_2(x) + \cdots = f(x).$$

This together with (6) shows that

$$f \simeq g_1 + g_2 + \cdots$$

which proves that f is integrable. Moreover, since

$$\int f_n = \int f_{n1} + \int f_{n2} + \cdots \quad (n = 1, 2, \ldots)$$

we have

$$\int f = \int g_1 + \int g_2 + \cdots = \int f_{11} + \int f_{12} + \cdots + \int f_{21} + \int f_{22} + \cdots = \int f_1 + \int f_2 + \cdots,$$

proving (2).

Note that in the above proof Theorem 4.6.6 has been used several times. In practice we often use the following corollary.

7.7.2 Corollary   *If $f_1, f_2, \ldots$ are integrable functions and*

$$\int |f_1| + \int |f_2| + \cdots < \infty,$$

then there exists an integrable function f such that

$$f \simeq f_1 + f_2 + \cdots.$$

Exercises 7.7

1. Prove Corollary 7.7.2.
2. Let f be the characteristic function of the set of all rational numbers. Show that f is integrable and find $\int f$.
3. Prove that any function that vanishes outside a countable number of points is integrable and that its integral is 0.

7.8. NULL FUNCTIONS AND NULL SETS

7.8.1 Definition   A function f is called a *null function* if it is integrable and if $\int |f| = 0$. Similarly, a set is called a *null set* if its characteristic function is a null function.

The role of null functions in the theory of the integral is like that of the number 0 in the theory of real numbers. The main difference is that the number 0 is unique whereas the class of null functions is infinite. The function that is identically equal to 0 is a null function. Also any function that vanishes outside a single point, a finite number of points, or a countable number of points, is a null function.

We start with the following simple observation.

7.8.2 Theorem   *A function f is a null function if and only if it vanishes everywhere except for a null set.*

Proof.   Let f be a null function and let Z be the set of all points x such that $f(x) \neq 0$. If g is the characteristic function of Z, then $g \simeq f + f + \cdots$. Since $\int f = 0$, we have $\int g = 0$. This shows that Z is a null set.

Assume now that Z is a null set and g is its characteristic function. If $f = 0$ outside Z, then $f \simeq g + g + \cdots$ and hence $\int f = 0$, which implies that f is a null function.

7.8.3 Theorem   *The sum of two null functions is a null function, and the product of a number and a null function is a null function. Moreover, the modulus of a null function is a null function.*

Proof.   If f and g are null functions, then

$$\int |f + g| \leq \int (|f| + |g|) = \int |f| + \int |g| = 0 + 0 = 0.$$

Thus the sum $f + g$ is a null function. For the product the assertion follows from the equality $\int |\lambda f| = |\lambda| \int |f|$, and for the modulus from the obvious equality $||f|| = |f|$.

7.8.4 Definition   Two integrable functions f and g are called *equivalent* if there exists a null function h such that $f = g + h$. If f and g are equivalent we write

$$f = g \text{ a.e.}$$

which is read "f equals g almost everywhere."

The notation $f = g$ a.e. is very popular and is convenient because it suggests that equality almost everywhere has properties similar to those of ordinary equality, namely:

$$f = f \text{ a.e.}$$
$$\text{If } f = g \text{ a.e.,} \quad \text{then} \quad g = f \text{ a.e.}$$
$$\text{If } f = g \text{ a.e.} \quad \text{and} \quad g = h \text{ a.e.,} \quad \text{then} \quad f = h \text{ a.e.}$$

Note that f *is a null function if and only if* $f = 0$ *a.e.*

Exercises 7.8

1.   Prove that $f = g$ a.e. if and only if $\int |f - g| = 0$.
2.   Prove the following:

   (a) $f = f$ a.e.

   (b)   If $f = g$ a.e., then $g = f$ a.e.

   (c)   If $f = g$ a.e. and $g = h$ a.e., then $f = g$ a.e.

3. Prove that if f is a null function and $|g| \leq f$, then g is a null function.

4. Prove that every subset of a null set is a null set.

5. Let f be an arbitrary function and let g be a null function. Prove that the product fg is a null function.

7.9 NORM CONVERGENCE

We have already considered two types of convergence of sequences of functions: pointwise convergence (Section 4.10) and uniform convergence (Section 4.11). In this and the following section we introduce two new types of convergence that are more natural and more important in the theory of the integral.

7.9.1 Definition We say that a sequence of integrable functions $f_1, f_2, \ldots$ converges to an integrable function f *in norm,* and we write $f_n \to f$ i.n., if $\|f_n - f\| \to 0$ as $n \to \infty$.

The norm convergence has similar algebraic properties to those of pointwise or uniform convergence:

7.9.2 *If $f_n \to f$ i.n. and $g_n \to g$ i.n., then*

$$f_n + g_n \to f + g \ i.n.,$$
$$f_n - g_n \to f - g \ i.n.,$$
$$\lambda f_n \to \lambda f \ i.n. \qquad (\lambda \in \mathbb{R}),$$
$$|f_n| \to |f| \ i.n..$$

The name norm convergence originates from the integral $\int |f|$ which, in the theory of integrable functions, is called the norm of f.

7.9.3 Theorem *If $f_n \to f$ i.n., then $\int f_n \to \int f$.*

Proof follows directly from the inequality

$$\left| \int f_n - \int f \right| \leq \int |f_n - f|.$$

A series of integrable functions $\left\{ \sum_{k=1}^{n} f_k \right\}$ is said to *converge to f in norm* if the sequence of partial sums $f_1 + \cdots + f_n$ converges to f in norm. In such a case we write $f_1 + f_2 + \cdots = f$ i.n.

7.9.4 Theorem *If*

$$f \simeq f_1 + f_2 + \cdots,$$

then the series converges to f in norm, that is,

$$f_1 + f_2 + \cdots = f \ i.n.$$

Proof.   Given any $\varepsilon > 0$ we choose an index n_0 such that

$$\int |f_{n+1}| + \int |f_{n+2}| + \cdots < \varepsilon \text{ for all } n > n_0.$$

Since

$$f - f_1 - \cdots - f_n \simeq f_{n+1} + f_{n+2} + \cdots$$

we have

$$\int |f - f_1 - \cdots - f_n| \le \int |f_{n+1}| + \int |f_{n+2}| + \cdots < \varepsilon$$

for all $n > n_0$, which proves the theorem.

7.9.5 Theorem   *If $f_n \to f$ i.n. and $f = g$ a.e., then $f_n \to g$ i.n. Conversely, if $f_n \to f$ i.n. and $f_n \to g$ i.n., then $f = g$ a.e. In other words, the limit of a sequence convergent in norm is defined up to a null function.*

Proof.   Let $f_n \to f$ i.n. and $g = f + h$ for some null function h. Then

$$\int |f_n - g)| = \int |f_n - (f + h)| \le \int |f_n - f| + \int |h|.$$

Since $\int |h| = 0$, we have $f_n \to g$ i.n.

Now, assume that $f_n \to f$ i.n. and $f_n \to g$ i.n. Then

$$\int |f - g| \le \int |f_n - f| + \int |f_n - g|.$$

Since the left side does not depend on n and the right side tends to 0, we obtain $\int |f - g| = 0$, which proves that $f = g$ a.e.

In the next theorem it is necessary to indicate the variable with respect to which the function is integrated. For this reason we write $\int f(t)dt$ instead of the usual $\int f$.

7.9.6 Theorem   *If f is an integrable function, then*

$$\lim_{h \to 0} \int |f(t + h) - f(t)|dt = 0.$$

Proof.   The assertion is obviously true if f is a brick function. Hence it can be easily extended for an arbitrary step function. Let $f \simeq f_1 + f_2 + \cdots$ be an arbitrary integrable function expanded into a series of step functions f_n. Given any $\varepsilon > 0$, there exists an index $n_0 \in \mathbb{N}$ such that

$$\sum_{k = n_o + 1}^{\infty} \int |f_k| < \frac{1}{2}\varepsilon.$$

Since the theorem is true of step functions and $f_1 + \cdots + f_{n_o}$ is a step function, we have

(1)
$$\lim_{h \to 0} \int \left| \sum_{k=1}^{n_o} (f_k(t+h) - f_k(t)) \right| dt = 0.$$

For any $h \in \mathbb{R}$, we have

$$\int |f(t+h) - f(t)| \, dt = \int \left| \sum_{k=1}^{n_o} (f_k(t+h) - f_k(t)) + \sum_{k=n_o+1}^{\infty} (f_k(t+h) - f_k(t)) \right| dt$$

$$\leq \int \left| \sum_{k=1}^{n_o} (f_k(t+h) - f_k(t)) \right| dt + \int \left| \sum_{k=n_o+1}^{\infty} (f_k(t+h) - f_k(t)) \right| dt$$

$$\leq \int \left| \sum_{k=1}^{n_o} (f_k(t+h) - f_k(t)) \right| dt + \sum_{k=n_o+1}^{\infty} \int |f_k(t+h) - f_k(t)| \, dt$$

$$\leq \int \left| \sum_{k=1}^{n_o} (f_k(t+h) - f_k(t)) \right| dt + \sum_{k=n_o+1}^{\infty} \int |f_k(t+h)| \, dt + \sum_{k=n_o+1}^{\infty} \int |f_k(t)| \, dt$$

$$\leq \int \left| \sum_{k=1}^{n_o} (f_k(t+h) - f_k(t)) \right| dt + \frac{1}{2}\varepsilon + \frac{1}{2}\varepsilon.$$

Thus, in view of (1), we obtain

$$\lim_{h \to 0} \int |f(t+h) - f(t)| \, dt \leq \lim_{h \to 0} \int \left| \sum_{k=1}^{n_o} (f_k(t+h) - f_k(t)) \right| dt + \varepsilon = \varepsilon.$$

Hence the assertion follows.

Exercises 7.9

1. Prove that $\int |f + g| \leq \int |f| + \int |g|$ for any integrable functions f and g.
2. Prove 7.9.2.
3. Prove 7.9.6 for step functions.
4. Let f be an integrable function and let $f_h(x) = f(x + h)$. Prove that $\int f_h = \int f$.
5. True or false? If $\{f_n\}$ is a sequence of integrable functions convergent to 0 uniformly on $\mathbb{R}$, then $f_n \to 0$ i.n.
6. Let $\{f_n\}$ be a sequence of integrable functions vanishing outside a bounded interval $[a, b]$. Prove that if $\{f_n\}$ converges to 0 uniformly on $[a, b]$ then $f_n \to 0$ i.n.

7.10 CONVERGENCE ALMOST EVERYWHERE

We proved in Section 7.5 that, if $f \simeq \lambda_1 f_1 + \lambda_2 f_2 + \cdots$, then the series converges at every point outside a null set. This result can be easily generalized:

7.10.1 Theorem   *If the series $\left\{ \sum_{k=1}^{n} \int |f_k| \right\}$ converges, then the series $\left\{ \sum_{k=1}^{n} f_k(x) \right\}$ converges everywhere except for a null set.*

Proof.   Let Z be the set of all points where the series $\left\{ \sum_{k=1}^{n} f_k(x) \right\}$ does not converge absolutely and let g be the characteristic function of Z. Then $g \simeq f_1 - f_1 + f_2 - f_2 + \cdots$, and hence $\int g = 0$.

The type of convergence in the above theorem is called *convergence almost everywhere*.

7.10.2 Definition   We say that a sequence of functions f_n converges to a function f *almost everywhere* if $\lim_{n \to \infty} f_n(x) = f(x)$ for all x except a null set. We then write $f_n \to f$ a.e.

7.10.3 Theorem   *A limit of a sequence convergent almost everywhere is determined up to a null function, that is, if $f_n \to f$ a.e., then $f_n \to g$ a.e. if and only if $f = g$ a.e.*

To prove this theorem it is useful to note that *the union of two null sets is a null set*. In order to see this, denote by χ_X and χ_Y the characteristic functions of null sets X and Y, and by $\chi_{X \cup Y}$ the characteristic function of the union $X \cup Y$. Then $\chi_{X \cup Y} \leq \chi_X + \chi_Y$ and hence $\int \chi_{X \cup Y} \leq \int \chi_X + \int \chi_Y = 0$.

Proof of 7.10.3.   Let $f_n \to f$ a.e., that is, $f_n \to f$ outside a null set X. If $f = g$ outside a null set Y, then $f_n \to g$ outside $X \cup Y$. Thus $f_n \to g$ a.e.

Now assume that $f_n \to f$ outside a null set X and $f_n \to g$ outside a null set Y. Then $f = g$ outside $X \cup Y$, which completes the proof.

The following auxiliary theorem will be useful in proofs of further properties of convergence almost everywhere.

7.10.4 Theorem   *A sequence $\{f_n\}$ converges to f almost everywhere if and only if there exist null functions h_n such that the sequence $\{f_n + h_n\}$ converges at every point.*

Proof.   Assume first that $f_n \to f$ a.e. Denote by Z the set of all points where the sequence $\{f_n(x)\}$ does not converge to $f(x)$. Let $h_n(x) = f(x) - f_n(x)$ for every $x \in Z$ and $h_n(x) = 0$ outside Z. Then the h_n are null functions and the sequence of functions $f_n + h_n$ converges to f everywhere.

Assume now, conversely, that $g_n = f_n + h_n$ and that the sequence $\{g_n\}$ converges to f at every point. If the h_n are null functions, then

$$\int |g_1 - f_1| + \int |g_2 - f_2| + \cdots = 0.$$

Hence, by 7.10.1, the series $\left\{ \sum_{k=1}^{n} (g_k - f_k) \right\}$ converges everywhere except on a null set Z. Therefore $g_n - f_n \to 0$ outside Z and, consequently, $f_n = g_n - (g_n - f_n) \to f$ outside Z, which completes the proof.

In view of Theorem 7.10.4, the following properties of convergence almost everywhere can be easily derived from the analogous properties of pointwise convergence.

7.10.5 Theorem   *If $f_n \to f$ a.e. and $g_n \to g$ a.e., then*

$$f_n + g_n \to f + g \ a.e.;$$
$$f_n - g_n \to f - g \ a.e.;$$
$$f_n g_n \to fg \ a.e.;$$
$$|f_n| \to |f| \ a.e.$$

7.10.6 Theorem   *If $f \simeq f_1 + f_2 + \cdots$, then the series converges to f almost everywhere, that is, $f_1 + f_2 + \cdots = f$ a.e.*

Proof.   In Section 7.4 we proved that the series converges absolutely at every point except on a null set Z. Clearly, the sum must be equal to f outside Z, which proves the theorem.

7.10.7 Theorem   *If $\int |f_1| + \int |f_2| + \cdots < \infty$, then*

$$f_1 + f_2 + \cdots = f \ i.n. \quad \text{if and only if} \quad f_1 + f_2 + \cdots = f \ a.e.$$

Proof.   Let g be any function such that $g \simeq f_1 + f_2 + \cdots$ Then both $f_1 + f_2 + \cdots = g$ i.n. and $f_1 + f_2 + \cdots = g$ a.e. hold by 7.9.4 and 7.10.6. If $f_1 + f_2 + \cdots = f$ i.n. then f and g differ by a null function from one another, by 7.9.5. Hence, by 7.10.3, we have $f_1 + f_2 + \cdots = f$ a.e. The proof of the converse is similar.

The above theorem describes a special case when convergence in norm implies convergence almost everywhere and conversely. As we see in the following two examples, in general neither implication holds.

7.10.8 Example   Consider the following sequence of functions:

$$f_n(x) = \begin{cases} n & \text{if } x \in [n, n+1), \\ 0 & \text{otherwise.} \end{cases}$$

For every $x \in \mathbb{R}$, $\lim_{n \to \infty} f_n(x) = 0$. Thus $f_n \to 0$ a.e. On the other hand, since $\int f_n = n$, the sequence is not convergent in norm.   ∎

7.10.9 Example   Consider the following sequence of functions:

$$f_n(x) = \begin{cases} 1 & \text{if } x \in \left[\dfrac{n}{2^{k-1}} - 1, \dfrac{n+1}{2^{k-1}} - 1 \right] \text{and } 2^{k-1} \le n < 2^k \text{ for some } k \in \mathbb{N}, \\ 0 & \text{otherwise.} \end{cases}$$

Although functions $f_1, f_2, f_3, \ldots$, are defined in a complicated manner, the idea is simple, as one can see in Fig. 7.8. Since

$$\int f_n = \frac{1}{2^k}$$

for every n such that $2^{k-1} \le n < 2^k$, we have $\|f_n\| = \int f_n \to 0$ as $n \to \infty$. Thus $f_n \to 0$ i.n. On the other hand, for every $x \in [0, 1]$, $f_n(x) = 0$ for infinitely many $n \in \mathbb{N}$, and $f_n(x) = 1$ for infinitely many $n \in \mathbb{N}$. Consequently, the sequence $\{f_n(x)\}$ does not converge at any $x \in [0, 1]$. Since $[0, 1]$ is not a null set, $f_n \not\to 0$ a.e. ∎

Exercises 7.10

1. Prove that a countable union of null sets is a null set.

2. Prove 7.10.5.

3. True or false? If $f_n \to f$ a.e., then $\int f_n \to \int f$.

4. Modify Example 7.10.8 so that all the functions in the sequence $\{f_n\}$ vanish outside $[0, 1]$ and yet $f_n \to 0$ a.e. and $\int f_n \to \infty$.

5. Modify Example 7.10.9 so that the sequence $\{f_n(x)\}$ does not converge at any $x \in \mathbb{R}$ and yet still $f_n \to 0$ i.n.

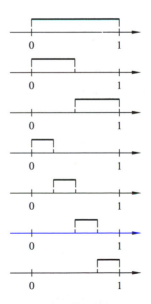

Fig. 7.8 A sequence convergent in norm but divergent everywhere on $[0, 1]$.

7.11 THEOREMS OF RIESZ

The following two theorems are attributed to the Hungarian mathematician Frigyes Riesz (1880–1956). They are existence theorems. The first of them asserts the existence of integrable functions that are limits in norm of sequences satisfying a certain condition (called Cauchy's condition). The second asserts the existence of subsequences convergent almost everywhere for sequences convergent in norm.

7.11.1 Theorem *Let* $\{f_n\}$ *be a sequence of integrable functions. If*

$$f_{p_{n+1}} - f_{p_n} \to 0 \text{ i.n. as } n \to \infty$$

,

for every increasing sequence of indices $p_1, p_2, \ldots,$ *then there is an integrable function f such that $f_n \to f$ i.n.*

Proof.   Let $\{\varepsilon_n\}$ be a sequence of positive numbers such that $\sum\limits_{k=1}^{\infty} \varepsilon_k < \infty$, and let $\{p_n\}$ be an increasing sequence of natural numbers such that

(1) $$\int |f_k - f_m| < \varepsilon_n \quad \text{for all } k, m > p_n.$$

(Why does such a sequence $\{p_n\}$ always exist?) Then

$$\int |f_{p_1}| + \int |f_{p_2} - f_{p_1}| + \int |f_{p_3} - f_{p_2}| + \cdots < \infty.$$

Hence, by 7.7.2, there exists a function f such that

$$f \simeq f_{p_1} + (f_{p_2} - f_{p_1}) + (f_{p_3} - f_{p_2}) + \cdots.$$

Thus, by 7.9.4, we have $\int |f_{p_n} - f| \to 0$. Since, by (1),

$$\int |f_k - f_{p_m}| < \varepsilon_n \quad \text{for all } k, p_m > p_n,$$

we have

$$\int |f_k - f| \le \int |f_k - f_{p_m}| + \int |f_{p_m} - f|.$$

By letting $m \to \infty$ we obtain

$$\int |f_k - f| \le \varepsilon_n \quad \text{for all } k > p_n.$$

Since $\varepsilon_n \to 0$, the above implies that the sequence $\{f_n\}$ converges to f in norm.

Example 7.10.9 shows that convergence in norm does not imply convergence almost everywhere. On the other hand we have the following important theorem.

7.11.2 Theorem   *If $f_n \to f$ i.n., then there is a subsequence $\{f_{p_n}\}$ of $\{f_n\}$ such that $f_{p_n} \to f$ a.e.*

Proof.   Let $\{\varepsilon_n\}$ be a sequence of positive numbers such that $\sum_{k=1}^{\infty} \varepsilon_k < \infty$. Since $\int |f_n - f| \to 0$, there exists an increasing sequence of natural numbers $p_1, p_2, \ldots$ such that

$$\int |f_{p_n} - f| < \varepsilon_n \quad \text{for all } n \in \mathbb{N}.$$

Then

$$\int |f_{p_{n+1}} - f_{p_n}| \leq \int |f_{p_{n+1}} - f| + \int |f - f_{p_n}| < \varepsilon_{n+1} + \varepsilon_n,$$

and consequently

$$\int |f_{p_1}| + \int |f_{p_2} - f_{p_1}| + \int |f_{p_3} - f_{p_2}| + \cdots < \infty.$$

Hence, there is a function g such that

$$g \simeq f_{p_1} + (f_{p_2} - f_{p_1}) + (f_{p_3} - f_{p_2}) + \cdots.$$

This, by 7.10.6, implies $f_{p_n} \to g$ a.e., and also $f_{p_n} \to g$ i.n., by 7.10.7. In view of 7.9.5 and the fact that $f_n \to f$ i.n., we have then $f = g$ a.e. Therefore, by 7.10.3, $f_{p_n} \to f$ a.e., completing the proof.

Exercises 7.11

1. Complete the proof of Theorem 7.11.1 by proving the existence of a sequence $\{p_n\}$ satisfying (1).

2. Find a subsequence $\{f_{p_n}\}$ of the sequence $\{f_n\}$ defined in Example 7.10.9 such that $f_{p_n} \to 0$ a.e.

3. True or false: If $f_n \to f$ a.e., then there is a subsequence $\{f_{p_n}\}$ of $\{f_n\}$ such that $f_{p_n} \to f$ i.n.

7.12 MONOTONE CONVERGENCE THEOREM

One the main reasons of the importance of convergence in norm is that it implies convergence of integrals, that is, $f_n \to f$ i.n. implies $\int f_n \to \int f$. Convergence almost everywhere does not have that property. On the other hand, for a particular sequence $\{f_n\}$ it is usually much easier to verify convergence almost everywhere than convergence in norm. Theorems giving conditions that imply convergence in norm are of great importance. In this and the following section we prove two such theorems. They belong to the most important theorems in the theory of the Lebesgue integral. The first one is due to the Italian mathematician Beppo Levi (1875–1961).

7.12.1 Theorem (monotone convergence theorem)   *If $\{f_n\}$ is a monotone sequence of integrable functions and $\int|f_n| \le M$, for some M and all $n \in \mathbb{N}$, then there exists an integrable function f such that $f_n \to f$ i.n. and $f_n \to f$ a.e.*

Proof.   We may assume that the sequence is non-negative and nondecreasing (otherwise we would consider $\{f_n - f_1\}$ or $\{f_1 - f_n\}$). Then

$$\int |f_1| + \int |f_2 - f_1| + \cdots + \int |f_n - f_{n-1}| = \int |f_n| \le M.$$

Hence, by letting $n \to \infty$, we obtain

$$\int |f_1| + \int |f_2 - f_1| + \cdots \le M < \infty.$$

By 7.7.2, there is an integrable function f such that

$$f \simeq f_1 + (f_2 - f_1) + \cdots.$$

Then $f = f_1 + (f_2 - f_1) + \cdots$ i.n. and $f = f_1 + (f_2 - f_1) + \cdots$ a.e., by 7.9.4 and 7.10.6, which proves the theorem.

Exercises 7.12

1. Find an example of a sequence of functions that shows that Theorem 7.12.1 does not hold if the assumption of monotonicity is dropped.

2. Find an example of a sequence of functions which shows that Theorem 7.12.1 does not hold if the assumption that $\int|f_n| \le M$ is dropped.

7.13 DOMINATED CONVERGENCE THEOREM

We are finally ready to prove the famous dominated convergence theorem, proved by Lebesgue in his doctoral dissertation in 1902. Without much exaggeration, one can say that this theorem was the main goal of the development of the first twelve sections of this chapter. This section closes the presentation of the basic theory of the Lebesgue integral in this book. The following sections will consist of applications and complements of the developed theory.

We shall use the following notation:

$$f_1 \vee \cdots \vee f_n = \max(|f_1|, \ldots, |f_n|).$$

If $f_1, \ldots, f_n$ are integrable, then so is $f_1 \vee \ldots \vee f_n$. This follows by induction, from the equality

$$f \vee g = \frac{1}{2}(|f| + |g| + ||f| - |g||),$$

which can be checked easily.

7.13.1 Theorem (dominated convergence theorem) *If a sequence of integrable functions $\{f_n\}$ converges almost everywhere to f and is bounded by an integrable function g (i.e., $|f_n| \le g$ for every $n \in \mathbb{N}$), then f is integrable and $f_n \to f$ i.n., and hence also $\int f_n \to \int f$.*

Proof. For $m, n = 1, 2, \ldots$, define

$$h_{n,m} = f_n \vee \cdots \vee f_{n+m}.$$

Then, for every fixed $n \in \mathbb{N}$, the sequence $\{h_{n,1}, h_{n,2}, \ldots\}$ is nondecreasing and, since

$$\left| \int h_{n,m} \right| = \int h_{n,m} \le \int g < \infty,$$

there is an integrable function h_n such that $h_{n,m} \to h_n$ a.e. as $m \to \infty$. Note that the sequence $\{h_n\}$ is nonincreasing and $0 \le h_n$ for all $n \in \mathbb{N}$. Thus it converges to a function h at every point and, by the monotone convergence theorem, h is integrable and $h_n \to h$ i.n. Now we will consider two cases.

Case 1. First suppose $f = 0$. Then $f_n \to 0$ a.e., and therefore $h_n \to 0$ a.e. Since the sequence converges in norm, we obtain $h_n \to 0$ i.n. Hence

$$\int |f_n| \le \int h_n \to 0,$$

which proves the theorem in the first case.

Case 2. When f is an arbitrary function, then for every increasing sequence of positive integers $\{p_n\}$ we have

$$g_n = f_{p_{n+1}} - f_{p_n} \to 0 \quad \text{a.e.}$$

and $|g_n| \le 2g$ for every $n \in \mathbb{N}$. Therefore, the sequence $\{g_n\}$ satisfies the assumptions of Case 1, and consequently $g_n \to 0$ i.n. Hence, by the first Riesz theorem (7.11.1), the sequence $\{f_n\}$ converges in norm to some integrable function f^*. On the other hand, by the second Riesz theorem (7.11.2), there exists an increasing sequence of positive integers q_n such that $f_{q_n} \to f^*$ a.e. But $f_{q_n} \to f$ a.e., and thus $f^* = f$ a.e. In view of Theorem 7.9.5, this implies that $f_n \to f$ i.n. This completes the proof.

Exercises 7.13

1. Prove that if $f_1, \ldots, f_n$ are integrable then so is $f_1 \vee \cdots \vee f_n$.

2. Find an example of a sequence of functions that shows that Theorem 7.13.1 does not hold if the assumption $|f_n| \le g$ is dropped.

3. Let f be an integrable function. Define

$$f_n(x) = \begin{cases} f(x) & \text{if } |x| \le n, \\ 0 & \text{otherwise.} \end{cases}$$

Show that $f_n \to f$ i.n.

4.   In the proof of Theorem 7.13.1 we claimed that the sequence $\{h_n\}$ is nonincreasing. Justify that claim.

7.14 INTEGRAL OVER AN INTERVAL

By $\int_a^b f$ we mean the integral $\int f\chi$ where χ is the characteristic function of the interval (a, b) (a and b may be finite or infinite). This definition makes sense whenever the product $f\chi$ is integrable. We assume that the product vanishes outside the interval (a, b) no matter what is the value of f or whether f is defined there.

Instead of the interval (a, b) we may take intervals the $[a, b)$, $[a, b]$, or $(a, b]$, because the change of the integrand at points a and b does not affect the existence or the value of the integral. If f is an integrable function, then $\int_a^b f$ exists for every a and b such that $a < b$. Indeed, if $f \simeq f_1 + f_2 + \cdots$, then $f\chi \simeq f_1\chi + f_2\chi + \cdots$. Similarly, if f is integrable over (A, B) and $A < a < b < B$, then f is integrable over (a, b).

7.14.1 Theorem   *If $a < b < c$, then the integrability of f over (a, b) and (b, c) implies the integrability of f over (a, c), and*

$$\int_a^c f = \int_a^b f + \int_b^c f.$$

Proof.   Let χ_1, χ_2, and χ_3 be the characteristic functions of (a, b), (b, c), and (a, c), respectively. Then $f\chi_3 = f\chi_2 + f\chi_1$ a.e. (everywhere except at b) and hence

$$\int f\chi_3 = \int f\chi_1 + \int f\chi_2.$$

It is convenient to define

$$\int_a^a f = 0 \quad \text{and} \quad \int_b^a f = -\int_a^b f.$$

Then Theorem 7.14.1 holds for all a, b, and c.

From the definition of the integral over an interval it follows immediately that

$$\int_a^b \lambda f = \lambda \int_a^b f, \qquad \lambda \in \mathbb{R}.$$

The important dominated convergence theorem applies also to any interval:

7.14.2 Theorem   *If a sequence $\{f_n\}$ of functions integrable over an interval (a, b) converges to f almost everywhere on (a, b) and is bounded by an integrable* *function on* (a, b), *then* f *is integrable over* (a, b) *and* $\int_a^b |f_n - f| \to 0$, *and hence also* $\int_a^b f_n \to \int_a^b f$.

Proof. We assume that f, all the f_n, and the bounding function vanish outside (a, b), and then we apply Theorem 7.13.1.

7.14.3 Theorem *Every continuous function* f *on a bounded closed interval* $[a, b]$ *is integrable over that interval.*

Proof. For every $n \in \mathbb{N}$ define

$$f_n(x) = f\left(a + k\frac{b - a}{2^n}\right)$$

if $\quad x \in \left[a + k\dfrac{b - a}{2^n}, a + (k + 1)\dfrac{b - a}{2^n}\right)$ and $\quad k = 0, \dots, 2^n - 1$; and

$$f_n(x) = f(b) \quad \text{if} \quad x = b$$

(see Fig. 7.9). We shall show that $\lim_{n \to \infty} f_n(x) = f(x)$ for every $x \in [a, b]$. Suppose that for some $x_0 \in [a, b]$ the sequence $\{f_n(x_0)\}$ does not converge to $f(x_0)$. Then there exist $\varepsilon > 0$ and an increasing sequence of indices p_n such that

$$(1) \qquad |f_{p_n}(x_0) - f(x_0)| > \varepsilon \quad \text{for all } n \in \mathbb{N}.$$

Since f is continuous, there exists $\delta > 0$ such that

$$(2) \qquad |f(x) - f(x_0)| < \varepsilon \quad \text{whenever} \quad |x - x_0| < \delta.$$

Now, for some $k, m \in \mathbb{N}$, we have

$$(3) \qquad x_0 \in \left[a + k\frac{ba}{2^{p_m}}, \ a + (k + 1)\frac{b - a}{2^{p_m}}\right) \subset (x_0 - \delta, x_0 + \delta),$$

and then

$$f_{p_m}(x_0) = f(a + k\frac{b - a}{2^{p_m}}).$$

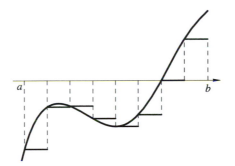

Fig 7.9 Approximation of a continuous
function by a step function.

Hence, by (2),

$$|f_{p_m}(x_0) - f(x_0)| = |f(a + k\frac{b-a}{2^{P_m}}) - f(x_0)| < \varepsilon,$$

which contradicts (1). This shows that $\lim_{n\to\infty} f_n(x) = f(x)$ for every $x \in [a, b]$. Since every continuous function on a closed interval is bounded, there exists a constant M such that $|f_n| < M$ on $[a, b]$ for all $n \in \mathbb{N}$. Therefore, by Theorem 7.14.2, f is integrable over $[a, b]$.

In Theorem 7.14.3 the closed interval $[a, b]$ cannot be replaced by the open interval (a, b); see Section 7.16.

Exercises 7.14

1. True or false? If f is integrable over $[-n, n]$ for all $n \in \mathbb{N}$, then f is integrable over $\mathbb{R}$.

2. A function $f : \mathbb{R} \to \mathbb{R}$ is called *locally integrable* if the product $f\chi_{[-n,n]}$ is an integrable function for every $n \in N$.

    (a) Prove that integrable functions are locally integrable. Is the converse true?

    (b) Prove that if f and g are locally integrable, then $f + g$ and $\lambda f (\lambda \in \mathbb{R})$ are locally integrable.

    (c) Prove that if a sequence of locally integrable functions $\{f_n\}$ converges almost everywhere to a function f and $|f_n| \le h$ for every $n \in \mathbb{N}$, where h is a locally integrable function, then f is locally integrable.

3. A set $S \subset \mathbb{R}$ is called *measurable* if χ_S is a locally integrable function. The family of all measurable subsets of $\mathbb{R}$ will be denoted by $\mathcal{M}$. The *measure* of a measurable set S, denoted by $\mu(S)$, is defined as

    $$\mu(S) = \begin{cases} \int \chi_S & \text{if } \chi_S \text{ is integrable} \\ \infty & \text{otherwise} \end{cases}$$

    (a) Prove that the empty set, a set of $\mathbb{Q}$ a single element, and $\mathbb{R}$ are measurable.

    (b) Prove that $[a, b], (a, b), [a, b), (a, b] \in \mathcal{M}$ for any $a < b$.

    (c) Prove that if $A, B \in \mathcal{M}$, then $A \cup B \in \mathcal{M}$ and $A \cap B \in \mathcal{M}$.

    (d) Prove that if S is measurable, then the complement of S is measurable.

    (e) Prove that if $A, B \in \mathcal{M}$ and $A \subset B$, then $\mu(A) \le \mu(B)$.

    (f) Prove that if $A \in \mathcal{M}$, $\mu(A) = 0$, and $B \subset A$, then $B \in \mathcal{M}$ and $\mu(B) = 0$.

    (g) Prove that if $A, B \in \mathcal{M}$ and A and B are disjoint, then $\mu(A \cup B) = \mu(A) + \mu(B)$. (We define $a + \infty = \infty$ and $\infty + \infty = \infty$).

7.15 INTEGRATION AND DIFFERENTIATION

If f is integrable on (a, b), then for any two points α and β in (a, b) the integral $\int_\alpha^\beta f$ exists. Consequently, the formula $F(x) = \int_c^x f$, where c is an arbitrary point in (a, b), defines a function on (a, b).

7.15.1 Theorem   *Let* $-\infty \le a < c < b \le \infty$ *and let* f *be an integrable function over* (a, b). *Then the function*

$$F(x) = \int_c^x f, \qquad a < x < b,$$

is continuous on (a, b) *and has limits* $F(a+)$ *and* $F(b-)$.

Proof.   Let x be any fixed point such that $a \le x < b$ and let $\{x_n\}$ be a decreasing sequence convergent to x, that is, $x < x_{n+1} < x_n$ and $x_n \to x$. Denote by g_n the characteristic function of the interval (x, x_n). Then the sequence of products $f g_n$ is convergent to zero at every point of (a, b). Since $|f g_n| \le |f|$, we have

$$F(x_n) - F(x) = \int_x^{x_n} f = \int_a^b f g_n \to 0,$$

by the dominated convergence theorem (in the form 7.14.2). Thus F is right-hand continuous at $x \in (a, b)$ and has a right-hand limit at a. Similarly, we can prove left-hand continuity and the existence of $F(b-)$ by considering the characteristic functions of intervals (x_n, x) where $x_n < x_{n+1} < x$ and $x_n \to x$. Since F is right-hand and left-hand continuous it is continuous on (a, b) and the limits $F(a+)$ and $F(b-)$ exist.

By the *right-hand* or *left-hand derivative* of a function f at a point x_0 we mean the limit, if it exists,

$$\lim_{h \to 0} \frac{f(x_0 + h) - f(x)}{h}$$

where h approaches 0 from the right or from the left, respectively.

7.15.2 Theorem   *If a function* f *is integrable over* (a, b) *and is right-hand continuous at a point* $x_0 \in [a, b)$ *(or left-hand continuous at a point* $x_0 \in (a, b]$), *then the function* $F(x) = \int_c^x f$, $c \in (a, b)$, *has a right-hand (or left-hand) derivative at* x_0 *which is equal to* $f(x_0+)$ *(or* $f(x_0-)$).

Proof.   In view of full symmetry it suffices to prove the assertion for the right-hand limit. We have

(1)
$$\frac{F(x_0 + h) - F(x_0)}{h} = \frac{\int_c^{x_0+h} f - \int_c^{x_0} f}{h} = \frac{1}{h} \int_{x_0}^{x_0+h} f.$$

For arbitrary $\varepsilon > 0$ there is $\delta > 0$ such that

$$f(x_0) - \varepsilon < f(t) < f(x_0) + \varepsilon \quad \text{for all } t \in (x_0, x_0 + \delta),$$

since f is right-hand continuous at x_0. Integrating these inequalities from x_0 to

$x_0 + h$, we get

$$h\,(f(x_0) - \varepsilon) < \int_{x_0}^{x_0+h} f < h\,(f(x_0) + \varepsilon) \quad \text{for} \quad 0 < h < \varepsilon,$$

because $\int_\alpha^\beta m = (\beta - \alpha)m$. Hence, by (1), we have

$$f(x_0) - \varepsilon < \frac{F(x_0 + h) - F(x_0)}{h} < f(x_0) + \varepsilon \quad \text{for } 0 < h < \varepsilon.$$

This means that

$$\lim_{h \to 0+} \frac{F(x_0 + h) - F(x_0)}{h} = f(x_0).$$

From the above theorem we can easily obtain the following.

7.15.3 Theorem *Every continuous function has an antiderivative,*

and

7.15.4 Theorem *If f has an integrable and continuous derivative f' on (a, b), then, for every $c \in (a, b)$, the difference $f(x) - \int_c^x f'$ is a constant function.*

7.15.5 Theorem *If f is continuous and integrable over a bounded or unbounded interval (a, b) and has an antiderivative F that has limits $\lim_{x \to a+} F(x) = F(a)$ and $\lim_{x \to b-} F(x) = F(b)$, then*

$$\int_a^b f = F(b) - F(a).$$

Proof. By 7.15.2 the function $G(x) = \int_c^x f$, $c \in (a, b)$, has the derivative f at every point of (a, b). It is thus an antiderivative of f. Since all antiderivatives differ from one another by a constant, we have $F(x) + C = G(x)$ for all $x \in (a, b)$. Hence

$$\int_a^b f = \int_a^c f + \int_c^b f = \int_c^b f - \int_c^a f = G(b) - G(a) = F(b) - F(a).$$

The above theorem reduces the evaluation of many integrals to the search for antiderivatives. This has been known for a long time. It is said that it was already used by Barrow, the teacher of Newton. It is the most important and most magical theorem of calculus. For that reason it is often called the *fundamental theorem of calculus*.

In calculations we often write $\int_a^b f(x)dx$ instead of $\int_a^b f$. This is very convenient if the integrand f is an algebraic expression of x. Another advantage is that it shows

Fig. 7.10 $\int_0^\pi \sin x\, dx = 2$.

clearly with respect to which variable the integration is performed, which is especially important in the case of integrals with parameters. For example, the following two integrals are essentially different: $\int_0^1 x^\alpha\, dx$ and $\int_0^1 x^\alpha\, d\alpha$.

7.15.6 Examples

1. To calculate the integral $\int_0^\pi \sin x\, dx$ (Fig. 7.10) we first find an antiderivative

$$\int \sin x\, dx = -\cos x$$

and then use 7.15.5:

$$\int_0^\pi \sin x\, dx = -\cos \pi + \cos 0 = 1 + 1 = 2.$$

Such a procedure is often written as $\int_0^\pi \sin x\, dx = [-\cos x]_0^\pi = 1 + 1 = 2$.

2.
$$\int_{-a}^a \sqrt{a^2 - x^2}\, dx = \left[\frac{x}{2}\sqrt{a^2 - x^2} + \frac{a^2}{2}\arcsin \frac{x}{a}\right]_{-a}^a = \frac{a^2}{2}\frac{\pi}{2} + \frac{a^2}{2}\frac{\pi}{2} = \frac{a^2\pi}{2}.$$

Note that in this example we evaluated the area of a semicircle of radius a (see Fig. 7.11)

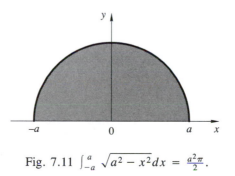

Fig. 7.11 $\int_{-a}^a \sqrt{a^2 - x^2}\, dx = \frac{a^2\pi}{2}$.

Exercises 7.15

1. Prove 7.15.3.

2. Prove 7.15.4.

3. Prove the *mean value theorem for integrals:*
    If f is a continuous function on $[a, b]$, *then there exists* $c \in (a, b)$ *such that*

    $$\int_a^b f = f(c)(b - a).$$

4. Prove that if f and g are continuous functions on $[a, b]$ and $\int_a^b f = \int_a^b g$, then there exists $c \in [a, b]$ such that $f(c) = g(c)$.

5. Let f be a continuous function on $[a, b]$. Define $g(x) = \int_x^b f$ for $x \in [a, b]$. Prove that g is differentiable. Express the derivative of g in terms of f. Caution: The points a and b require special treatment.

6. Let f be a continuous function on $[a, b]$. Prove that if $\int_a^x f = \int_x^b f$ for all $x \in [a, b]$ then $f(x) = 0$ for all $x \in [a, b]$.

7. Find an example of an unbounded continuous function on $\mathbb{R}$ that is integrable.

7.16 INTEGRABILITY OF $f(x) = x^\alpha$

We shall investigate the integrability of $f(x) = x^\alpha$ on the intervals $(0, 1)$ and $(1, \infty)$.

7.16.1 Theorem   *The function* $f(x) = x^\alpha$ *is integrable over* $(0, 1)$ *if and only if* $\alpha > -1$, *and in that case the integral is given by*

$$\int_0^1 x^\alpha dx = \frac{1}{\alpha + 1}.$$

The function $f(x) = x^\alpha$ *is integrable over* $(1, \infty)$ *if and only if* $\alpha < -1$, *and in that case the integral is given by*

$$\int_1^\infty x^\alpha dx = -\frac{1}{\alpha + 1}.$$

This theorem has a good intuitive interpretation. If $\alpha \geq 0$ (Fig. 7.12) then the function $f(x) = x^\alpha$ is continuous on the whole closed interval $[0, 1]$ and thus is integrable over that interval. If $\alpha < 0$ (Fig. 7.13) then the function is unbounded on the right neighborhood of the origin. The integral represents the shaded area under the curve. If the area is "thin enough," which depends on α, then it may happen to be finite. Otherwise it is infinite and the function is not integrable. Similarly, if we

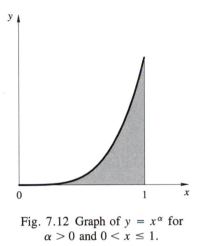

Fig. 7.12 Graph of $y = x^\alpha$ for
$\alpha > 0$ and $0 < x \leq 1$.

consider the function on the interval $(1, \infty)$, then the shaded area (Fig. 7.14) may be finite or infinite, depending on α.

Proof of 7.16.1. To prove the first part of the theorem, let us consider the functions

(1) $$f_n(x) = \begin{cases} x^\alpha & \text{for } x \in (\tfrac{1}{n}, 1) \\ 0 & \text{elsewhere} \end{cases}$$

$n = 1, 2, \ldots$. This is a monotone sequence of integrable functions, and

(2). $$0 < \int_{\frac{1}{n}}^{1} x^\alpha \, dx = \left[\frac{x^{\alpha+1}}{\alpha + 1} \right]_{\frac{1}{n}}^{1} = \frac{1}{\alpha + 1} \left(1 - n^{-\alpha-1} \right)$$

Fig. 7.13 Graph of
$y = x^\alpha$ for $\alpha < 0$
and $0 < x \leq 1$.

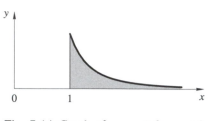

Fig. 7.14 Graph of $y = x^\alpha$ for $\alpha < 0$
and $1 < x \leq \infty$.

If $-1 < \alpha < 0$, the last expression is less than $1/\alpha + 1$ and hence the limit

(3)
$$f(x) = \begin{cases} x^\alpha & \text{for } x \in (0, 1) \\ 0 & \text{elsewhere} \end{cases}$$

is an integrable function, by the monotone convergence theorem. In other words, the function $f(x) = x^\alpha$ is integrable over $(0, 1)$ for every $\alpha \in (-1, 0)$. The value of the integral is

$$\int_0^1 x^\alpha dx = \lim_{n \to \infty} \frac{1}{\alpha + 1}\left(1 - n^{-\alpha-1}\right) = \frac{1}{\alpha + 1}.$$

Note that the integral has the same value $1/\alpha + 1$ for any $\alpha \geq 0$. Thus

$$\int_0^1 x^\alpha dx = \frac{1}{\alpha + 1} \quad \text{for all } \alpha > -1.$$

We still have to prove that the function $f(x) = x^\alpha$ is not integrable over $(0, 1)$ if $\alpha \leq -1$. Suppose, conversely, that it is integrable. Then its integral would be a positive number $\int_0^1 x^\alpha dx = \int f = M < \infty$, where f is defined as before by (3). Since $f_n \leq f$ (where the f_n are defined by (1)) we would have $\int f_n \leq \int f = M$. But for $\alpha < -1$, we have $\lim_{n \to \infty} \int f_n = \infty$. This contradiction shows that the function is not integrable for $\alpha < -1$. Finally, if $\alpha = -1$, we have

$$\int f_n = \int_{\frac{1}{n}}^1 \frac{dx}{x} = \ln n \to \infty$$

and as before we get a contradiction.

Let us now consider the functions

(4)
$$g_n(x) = \begin{cases} x^\alpha & \text{for } x \in (1, n) \\ 0 & \text{elsewhere} \end{cases}$$

where $n = 1, 2, \ldots$ This is a monotone sequence of integrable functions such that

$$0 < \int g_n = \int_1^n x^\alpha dx = \left[\frac{x^{\alpha+1}}{\alpha + 1} \right]_1^n = \frac{1}{\alpha + 1} \left(n^{\alpha+1} - 1 \right).$$

If $\alpha < -1$, then $0 < \int g_n < -\frac{1}{\alpha+1}$ and, by the monotone convergence theorem, the limit

(5) $$f(x) = \begin{cases} x^\alpha & \text{for } x \in (1, \infty) \\ 0 & \text{elsewhere} \end{cases}$$

is an integrable function on $(1, \infty)$ and the value of the integral is

$$\int_1^\infty x^\alpha dx = \lim_{n \to \infty} \frac{1}{\alpha + 1} \left(n^{\alpha+1} - 1 \right) = -\frac{1}{\alpha + 1}.$$

If $\alpha > -1$, then $\int g_n \to \infty$, which shows that g is not integrable, because $\int g_n \leq \int g$ for all $n \in \mathbb{N}$. Finally, if $\alpha = -1$, then

$$\int g_n = \int_1^n \frac{dx}{x} = \ln n \to \infty,$$

and again the limit function is not integrable. The proof is complete.

Exercises 7.16

1. Evaluate the following integrals:

(a) $\displaystyle \int_{-\infty}^{\infty} \frac{dx}{x^2 + 1}$,   (b) $\displaystyle \int_0^1 \ln x \, dx$.

7.17 FURTHER PRACTICAL CRITERIA OF INTEGRABILITY

The following simple criterion is often useful.

7.17.1 Theorem   *Let* $x_0 < x_1 < \cdots < x_n$. *If a function* f *is integrable over each of the intervals* (x_0, x_1), (x_1, x_2), $\ldots$, (x_{n-1}, x_n), *then* f *is integrable over the interval* (x_0, x_n) *and*

$$\int_{x_0}^{x_n} f = \int_{x_0}^{x_1} f + \cdots + \int_{x_{n-1}}^{x_n} f.$$

Proof. In the case $n = 2$ we return to Theorem 7.14.1. The general case follows by induction.

7.17.2 Theorem   *If a function f defined on an interval (a, b), bounded or unbounded, is integrable over every bounded interval (α, β) inside (a, b) (i.e., $a < \alpha < \beta < b$), and is bounded by an integrable function over the entire interval (a, b), (i.e. $|f| \leq g$ on (a, b) for some integrable function g), then f is integrable over (a, b).*

Proof.   Let

$$a < \cdots < \alpha_n < \alpha_{n-1} < \cdots < \alpha_1 < \beta_1 < \cdots < \beta_{n-1} < \beta_n < \cdots < b$$

and $\alpha_n \to a$ and $\beta_n \to b$. Then the functions

$$f_n(x) = \begin{cases} f(x) & \text{for } x \in (\alpha_n, \beta_n) \\ 0 & \text{elsewhere,} \end{cases}$$

(where $n = 1, 2, \ldots$), are integrable and $f_n(x) \to f(x)$ for every $x \in (a, b)$. Moreover, since $|f| \leq g$, then also $|f_n| \leq g$ for all $n \in \mathbb{N}$. Therefore, by the dominated convergence theorem, $f_n \to f$ i.n., which implies the integrability of f over (a, b).

7.17.3 Example   We shall show that the function $f(x) = x^\alpha e^{-x} (\alpha > -1)$ is integrable over $(0, \infty)$. First note that f is integrable over $(\varepsilon, 1)$ for any $0 < \varepsilon < 1$ and is bounded by the integrable function $g(x) = x^\alpha$. Thus it is integrable over the interval $(0, 1)$. To prove its integrability over $(1, \infty)$, observe that f is integrable over each of the intervals $(1, n)$, $(n = 2, 3, \ldots)$, because it is a continuous function. Let β be a number less than -1. Then there exists a number λ such that

$$0 < f(x) < \lambda x^\beta \quad \text{for all } x \in (1, \infty).$$

Since the function $h(x) = \lambda x^\beta$ is integrable over $(1, \infty)$, so is f. Since f is integrable over $(0, 1)$ and $(1, \infty)$ it is integrable over $(0, \infty)$, by 7.17.1. ■

7.17.4 Theorem   *Let g be a nondecreasing differentiable function defined on the interval (a, b), $-\infty \leq a < b \leq \infty$. Furthermore, we assume that g' is integrable over (a, b). Denote by $g(a+)$ and $g(b-)$ the right-hand limit at a and the left-hand limit at b, respectively (finite or infinite). If f is an integrable function over $(g(a+), g(b-))$, then the product $f(g(t))g'(t)$ is integrable over (a, b) and*

$$\int_{g(a+)}^{g(b-)} f(t) dt = \int_a^b f(g(t))g'(t) dt.$$

Proof.   If we take for granted that the product $f(g(t))g'(t)$ is integrable over (a, b) the proof is easy. Indeed, let F denote an antiderivative of f on the interval $(g(a+), g(b-))$. Then $F(g(t))$ is an antiderivative of $f(g(t))g'(t)$ and

$$\int_a^b f(g(t))g'(t) dt = F(g(b-)) - F(g(a+)) = \int_{g(a+)}^{g(b-)} f(t) dt.$$

It thus remains to prove that $f(g(t))g'(t)$ is integrable over (a, b). Let

$$f \simeq \lambda_1 f_1 + \lambda_2 f_2 + \cdots$$

where the f_n's are brick functions. There is no loss of generality if we assume that all brick functions f_n vanish outside of $(g(a+), g(b-))$. We will show that the function $\Phi(t) = f(g(t))g'(t)$ expands into

(1) $$\Phi \simeq \lambda_1 \Phi_1 + \lambda_2 \Phi_2 + \cdots$$

where

$$\Phi_n(t) = f_n(g(t))g'(t).$$

Let, for $n = 1, 2, \ldots, [a_n, b_n)$ be the support of f_n; that is, $f_n(t) = 1$ if $t \in [a_n, b_n)$, and $f_n(t) = 0$ if $t \notin [a_n, b_n)$. Since $g(a+) \leq a_n < b_n \leq g(b-)$ and g is continuous (being differentiable), there are numbers $\alpha_n < \beta_n$ such that $g(\alpha_n) = a_n$ and $g(\beta_n) = b_n$, for all $n \in \mathbb{N}$. Then

$$\int |\Phi_n| = \int_{\alpha_n}^{\beta_n} g'(t)dt = g(\beta_n) - g(\alpha_n) = b_n - a_n = \int f_n.$$

Since $|\lambda_1| \int f_1 + |\lambda_2| \int f_2 + \cdots < \infty$, we also have $\int |\lambda_1 \Phi_1| + \int |\lambda_2 \Phi_2| + \cdots < \infty$. Moreover, if the series $\left\{ \sum^n s_{k=1} \lambda_k \Phi_k(t) \right\}$ converges absolutely for some t, then it converges to $\Phi(t)$. This is trivially true if $g'(t) = 0$. If $g'(t) = k \neq 0$, then letting $g(t) = x$, we have

$$\lambda_1 \Phi_1(t) + \lambda_2 \Phi_2(t) + \cdots = (\lambda_1 f_1(x) + \lambda_2 f_2(x) + \cdots)k = f(x)k = \Phi(t),$$

the convergence being absolute. Thus (1) holds, which proves the integrability of $f(g(t))g'(t)$.

7.17.5 Corollary *If f is integrable over (a, b) and $\lambda \neq 0$, then*

$$\int_a^b f = \lambda \int_{a/\lambda}^{b/\lambda} f(\lambda t)dt.$$

Proof.   Take $g(t) = \lambda t$ and use 7.17.4.

7.17.6 Example (Euler's gamma function)   Euler's gamma function is defined by the integral

$$\Gamma(p) = \int_0^\infty x^{p-1} e^{-x} dx,$$

$p > 0$. Substituting $x = ty$ $(t > 0)$ we obtain

$$\Gamma(p) = t^p \int_0^\infty y^{p-1} e^{-ty} dy.$$

Exercises 7.17

1. Evaluate the following integrals:

   (a) $\displaystyle\int_0^3 \frac{dx}{9+x}$;

   (b) $\displaystyle\int_1^2 \frac{dx}{x^2+5x+4}$;

   (c) $\displaystyle\int_{-1}^1 \arccos x \, dx$;

   (d) $\displaystyle\int_0^1 \sqrt{x^2+1}\,dx$;

   (e) $\displaystyle\int_{-\infty}^\infty e^{-|x|}dx$;

   (f) $\displaystyle\int_0^3 \frac{dx}{\sqrt{3-x}}$.

2. Let

$$g(x) = \begin{cases} 1/\sqrt{x} & \text{for } 0 < |x| < 1 \\ 0 & \text{otherwise.} \end{cases}$$

   Show that g is integrable but g^2 is not.

3. Let $f(x) = \min\{1, 1/|x|\}$. Show that f is not integrable but f^2 is.

7.18 CAUCHY'S FORMULA

Let $[a, b]$ be an interval containing 0. For any function f integrable over $[a, b]$, the function $F_1(x) = \int_0^x f$ is continuous on $[a, b]$ and thus integrable. Consequently it can be integrated once more

$$F_2(x) = \int_0^x F_1 = \int_0^x \int_0^x f$$

and again

$$F_3(x) = \int_0^x F_2 = \int_0^x \int_0^x \int_0^x f$$

and so on. In this way obtain a sequence of functions defined on $[a, b]$:

$$F_1(x) = \int_0^x f,$$

$$F_n(x) = \int_0^x F_{n-1} = \underbrace{\int_0^x \cdots \int_0^x}_{n \text{ times}} f, \qquad n = 2, 3, 4, \ldots.$$

We are going to show that each of these iterated integrals can be expressed as a single integral.

7.18.1 Theorem (Cauchy's formula)   *Let $0 \in [a, b]$. For any function f, integrable over $[a, b]$, and any natural number $n \in \mathbb{N}$,*

(1)
$$\underbrace{\int_0^x \cdots \int_0^x}_{n \text{ times}} f = \frac{1}{(n-1)!} \int_0^x (x-t)^{n-1} f(t)\,dt$$

for $x \in [a, b]$.

Proof.   We shall use mathematical induction. Formula (1) is trivially true for $n = 1$. Assume that it holds for some $m \in \mathbb{N}$. First we show that

(2)
$$G(x) = \frac{1}{m!} \int_0^x (x-t)^m f(t)\,dt$$

is an antiderivative of

(3)
$$\frac{1}{(m-1)!} \int_0^x (x-t)^{m-1} f(t)\,dt,$$

$x \in [a, b]$. In fact

$$\frac{G(x+h) - G(x)}{h} = \frac{1}{hm!} \int_0^{x+h} (x+h-t)^m f(t)\,dt - \frac{1}{hm!} \int_0^x (x-t)^m f(t)\,dt$$

$$= \frac{1}{m!} \int_0^x \frac{(x+h-t)^m - (x-t)^m}{h} f(t)\,dt + \frac{1}{hm!} \int_x^{x+h} (x+h-t)^m f(t)\,dt.$$

Let $x \in [a, b]$ be fixed. For every $t \in [a, b]$, the difference quotient

$$\frac{(x+h-t)^m - (x-t)^m}{h}$$

converges to $m(x-t)^{m-1}$ as $h \to 0$. Since

$$\frac{(x+h-t)^m - (x-t)^m}{h} = m\xi^{m-1},$$

for some ξ between $x - t$ and $x + h - t$ there exists a constant M such that

$$\left| \frac{(x+h-t)^m - (x-t)^m}{h} \right| \le mM^{m-1}$$

for all $t \in [a, b]$ and all sufficiently small h. Therefore, by the dominated convergence theorem, we have

(5)
$$\lim_{h \to 0} \frac{1}{m!} \int_0^x \frac{(x+h-t)^m - (x-t)^m}{h} f(t)\,dt = \frac{1}{(m-1)!} \int_0^x (x-t)^{m-1} f(t)\,dt.$$

The second integral in (4) converges to zero, because $|x + h - t|^m \le |h|^m$ in the interval of integration, and consequently

$$\left| \frac{1}{hm!} \int_x^{x+h} (x + h - t)^m f(t)dt \right| \le \frac{|h|^{m-1}}{m!} \cdot \left| \int_x^{x+h} f(t)dt \right| \to 0$$

as $h \to 0$. This proves that

$$\lim_{h \to 0} \frac{G(x + h) - G(x)}{h} = \frac{1}{(m - 1)!} \int_0^x (x - t)^{m-1} f(t)dt.$$

Thus (2) is an antiderivative of (3). On the other hand,

$$\underbrace{\int_0^x \cdots \int_0^x}_{m + 1 \text{ times}} f$$

is also an antiderivative of (3). Since both functions are 0 at $x = 0$, they are equal. This shows that (1) holds for $m + 1$. Consequently, in view of the induction principle, (1) holds for every $n \in \mathbb{N}$.

Exercises 7.18

1. Justify the use of the dominated convergence theorem in (5).

7.19 TAYLOR'S FORMULA

Derivatives obtained by repeated differentiation are called *higher order derivatives*. The standard notation for higher order derivatives is somewhat inconsistent: $f'' = (f')'$, $f''' = (f'')'$, $f^{(4)} = (f''')'$, $f^{(5)} = (f^{(4)})'$, and so on.

7.19.1 Theorem   *Let $a < 0 < b$. If f has n continuous derivatives on (a, b), then*

$$(1) \qquad f(x) = f(0) + \frac{f'(0)}{1!}x + \frac{f''(0)}{2!}x^2 + \cdots + \frac{f^{(n-1)}(0)}{(n - 1)!}x^{n-1} + \underbrace{\int_0^x \cdots \int_0^x}_{n \text{ times}} f^{(n)}.$$

for $x \in (a, b)$.

Proof.   By Theorem 7.15.5 we may write

$$f^{(n-1)}(x) = f^{(n-1)}(0) + \int_0^x f^{(n)}(t)dt,$$

so that (1) holds for $n = 1$. For induction, assume that (1) holds for some n and all functions satisfying the conditions of the theorem. If we assume that $f^{(n+1)}$ is continuous on (a, b), we may replace f in (1) by f' and get

$$f'(x) = f'(0) + \frac{f''(0)}{1!}x + \cdots + \frac{f^{(n)}(0)}{(n-1)!}x^{n-1} + \underbrace{\int_0^x \cdots \int_0^x}_{n \text{ times}} f^{(n+1)}.$$

Integrating this equality from 0 to x we obtain

$$f(x) - f(0) = \frac{f'(0)}{1!}x + \frac{f''(0)}{2!} \cdots + \frac{f^{(n)}(0)}{n!}x^n + \underbrace{\int_0^x \cdots \int_0^x}_{n+1 \text{ times}} f^{(n+1)}.$$

This proves that the formula holds for all $n \in \mathbb{N}$.

The remainder

$$R_n = \underbrace{\int_0^x \cdots \int_0^x}_{n \text{ times}} f^{(n)}$$

can be transformed in several ways. For example:

Integral form: $\quad R_n = \frac{1}{(n-1)!} \int_0^x (x-t)^{n-1} f^{(n)}(t)\, dt.$

Lagrange's form: $\quad R_n = \frac{x^n}{n!} f^{(n)}(\xi)$, where ξ is a number between 0 and x that depends on x.

The integral form follows immediately from 7.18.1. To prove the Lagrange form we shall use the following theorem:

7.19.2 Theorem *If f is continuous on $[a, b]$ and g is non-negative and integrable over (a, b), then there exists $\xi \in (a, b)$ such that*

(2)
$$\int_a^b fg = f(\xi) \int_a^b g.$$

Proof. Denote by m and M the minimum and the maximum of f on $[a, b]$, respectively. Then

$$mg(t) \le f(t)g(t) \le Mg(t) \quad \text{for all } t \in (a, b).$$

Integrating these inequalities from a to b we get

$$m \int_a^b g \le \int_a^b fg \le M \int_a^b g.$$

On the other hand,

$$m \int_a^b g \leq f(t) \int_a^b g \leq M \int_a^b g,$$

for all $t \in [a, b]$. Since the function $f(t) \int_a^b g$ is continuous on $[a, b]$, there exists $\xi \in (a, b)$ for which (2) holds.

Applying the above theorem to the integral form of the remainder R_n we get the Lagrange remainder:

$$\frac{1}{(n-1)!} \int_0^x (x-t)^{n-1} f^{(n)}(t)dt = \frac{1}{(n-1)!} f^{(n)}(\xi) \int_0^x (x-t)^{n-1} dt = \frac{x^n}{n!} f^{(n)}(\xi)$$

for some ξ between 0 and x.

7.19.3 Examples

1. $e^x = 1 + \dfrac{x}{1!} + \dfrac{x^2}{2!} + \cdots + \dfrac{x^{n-1}}{(n-1)!} + \dfrac{x^n}{n!} e^\xi, \qquad |\xi| < |x|.$

2. $\ln(x+1) = \dfrac{x}{1!} - \dfrac{x^2}{2!} + \dfrac{x^3}{3!} - \cdots$

$$+ \frac{(-1)^{n-2}}{(n-1)!} x^{n-1} + \frac{(-1)^{n-1}}{n!} \frac{x^n}{(1+\xi)^n}, |\xi| < |x|.$$

3. $\sin x = \dfrac{x}{1!} - \dfrac{x^3}{3!} + \dfrac{x^5}{5!} - \cdots$

$$+ (-1)^{n-1} \frac{x^{2n-1}}{(2n-1)!} + \frac{x^{2n}}{(2n)!} \sin(\xi + n\pi), \qquad |\xi| < |x|,$$

because $(\sin x)^{(m)} = \sin(x + m\frac{\pi}{2})$.

Exercises 7.19

1.   Prove that $(\sin x)^{(m)} = \sin(x + m\frac{\pi}{2})$ for all $m \in \mathbb{N}$.
2.   Show that Theorem 7.19.2 does not hold if the assumption that g is non-negative is dropped.

7.20   INTEGRAL TEST FOR THE CONVERGENCE OF INFINITE SERIES

The theory of the integral gives a very convenient criterion for the convergence of infinite series.

7.20.1 Theorem (Cauchy's integral test)   *Let f be a positive decreasing function on $[1, \infty)$ tending to 0 as $x \to \infty$. Then the series $\{\sum_{k=1}^n f(k)\}$ converges if and only if the function f is integrable over $[1, \infty)$.*

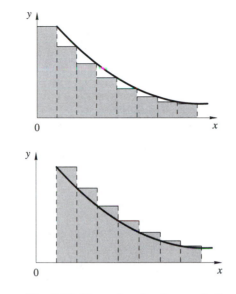

Fig. 7.15 Illustration for the proof of
Cauchy's integral test.

Proof.   The main idea of the proof is illustrated by Fig. 7.15. We have

(1)
$$\sum_{k=1}^{m} f(k+1) \leq \int_{1}^{m+1} f \leq \sum_{k=1}^{m} f(k),$$

which follows by induction from

$$f(k+1) \leq \int_{k}^{k+1} f \leq f(k).$$

From (1), by letting $m \to \infty$, we obtain

(2)
$$\sum_{k=2}^{\infty} f(k) \leq \int_{1}^{\infty} f \leq \sum_{k=1}^{\infty} f(k),$$

and the assertion follows easily.

7.20.2 Theorem   *The series*

$$\left\{ \sum_{k=1}^{n} \frac{1}{k^{\alpha}} \right\}$$

converges if and only if $\alpha > 1$.

Proof.   The assertion follows immediately from 7.20.1 and 7.16.1.

In particular, the harmonic series $\{\sum_{k=1}^{n} 1/k\}$ is divergent while the series $\{\sum_{k=1}^{n} 1/k^2\}$ is convergent. From (2) we obtain

$$1 + \frac{1}{2^2} + \frac{1}{3^2} + \cdots < \frac{4}{3},$$

because

$$\sum_{k=2}^{\infty} \frac{1}{k^2} \leq \int_1^{\infty} \frac{dx}{x^2} = \left[-\frac{1}{3} \frac{1}{x^3} \right]_1^{\infty} = \frac{1}{3}.$$

Exercises 7.20

1. Test for convergence

    (a) $\left\{ \sum_{k=2}^{n} \frac{1}{k(\ln\ k)^2} \right\}$;   (b) $\left\{ \sum_{k=2}^{n} \frac{1}{k \cdot \ln\ k \cdot \ln\ \ln\ k} \right\}.$

2. Consider the series

    $$\left\{ \sum_{k=2}^{n} \frac{1}{k^p (\ln k)^q} \right\}.$$

    (a) Prove that if $p > 1$ then the series converges for arbitrary q.
    (b) Prove that the series converges if $p = 1$ and $q > 1$.
    (c) Prove that the series diverges for arbitrary q if $p < 1$.
    (d) Prove that the series diverges if $p = 1$ and $q \leq 1$.

3. Complete the proof of Theorem 7.20.1.

7.21 EXAMPLES OF TRIGONOMETRIC SERIES

Even if we succeed in proving that a series is convergent, we do not know, in general, what its sum is. An approximate value can be found by adding a finite number of initial terms. This method can be used, for example, to approximate the sum of

$$1 + \frac{1}{2^2} + \frac{1}{3^2} + \cdots$$

But by no means would it lead to the discovery of the following magic connection between the series and the number π:

$$1 + \frac{1}{2^2} + \frac{1}{3^2} + \cdots = \frac{\pi^2}{6}.$$

The proof of this equality is much more difficult than the proof of the convergence of the series. We shall prove the equality using trigonometric series.

By a *trigonometric series* or *Fourier series* (Jean Baptiste Joseph Fourier, 1768–1830) we mean any series of the form

$$\left\{ \frac{1}{2}a_0 + \sum_{k=1}^{n}(a_k \cos kx + b_k \sin kx) \right\}.$$

We shall first consider some elementary cases. For example, we are going to show that

(1) $$\sum_{k=1}^{\infty} \frac{\sin kx}{k} = \frac{\pi - x}{2} \quad \text{for } x \in (0, 2\pi),$$

and that the series is uniformly convergent on every closed interval of the form $[\varepsilon, 2\pi - \varepsilon]$ contained in $[0, 2\pi]$.

In view of the equation

$$2 \cos nx \sin \tfrac{1}{2}x = -\sin(n - \tfrac{1}{2})x + \sin(n + \tfrac{1}{2})x$$

we have

$$2(\cos x + \cos 2x + \cdots + \cos nx)\sin \tfrac{1}{2}x$$
$$= (-\sin \tfrac{1}{2}x + \sin \tfrac{3}{2}x) + (-\sin \tfrac{3}{2}x + \sin \tfrac{5}{2}x) + \cdots$$
$$+ (-\sin(n - \tfrac{1}{2})x + \sin(n + \tfrac{1}{2})x)$$
$$= -\sin \tfrac{1}{2}x + \sin(n + \tfrac{1}{2})x,$$

and hence

$$\sum_{k=1}^{n} \cos kx = -\frac{1}{2} + \frac{\sin(n + \tfrac{1}{2})x}{2 \sin \tfrac{1}{2}x}.$$

Integrating from π to x we get

$$\sum_{k=1}^{n} \frac{\sin kx}{k} = \frac{\pi - x}{2} + I_n(x)$$

where

$$I_n(x) = \int_{\pi}^{x} \frac{\sin(n + \tfrac{1}{2})t}{2 \sin \tfrac{1}{2}t} dt.$$

Integration by parts yields

$$I_n(x) = -\frac{1}{2n + 1}\left(\frac{\cos(n + \tfrac{1}{2})x}{\sin \tfrac{1}{2}x} + \frac{1}{2}\int_{\pi}^{x} \frac{\cos \tfrac{1}{2}t \cos(n + \tfrac{1}{2})t}{(\sin \tfrac{1}{2}t)^2} dt \right);$$

hence

(2) $$|I_n(x)| < \frac{1}{2n + 1}\left(\frac{1}{\sin \tfrac{1}{2}\varepsilon} + \frac{1}{2}\int_{\varepsilon}^{\pi} \frac{dt}{(\sin \tfrac{1}{2}t)^2} \right) \quad \text{for } x \in [\varepsilon, 2\pi - \varepsilon].$$

This inequality proves that the sequence of functions I_n converges to 0 uniformly on $[\varepsilon, 2\pi - \varepsilon]$. Since ε is arbitrary, formula (1) follows.

For $x = 0$ the series in (1) is, as one can see immediately, convergent to 0, and not to $(\pi - x)/2$. Moreover, since all the terms of the series are periodic, series (1) converges to a periodic function whose graph is seen in Fig. 7.16.

Let

$$H(x) = -\sum_{k=1}^{\infty} \frac{\cos kx}{k^2}$$

(note that this series converges uniformly on $[0, 2\pi]$). Since

$$H'(x) = \left(-\sum_{k=1}^{\infty} \frac{\cos kx}{k^2}\right)' = -\sum_{k=1}^{\infty} \left(\frac{\cos kx}{k^2}\right)' = \sum_{k=1}^{\infty} \frac{\sin kx}{k} = \frac{\pi - x}{2},$$

we have

$$H(x) = -\frac{(\pi - x)^2}{4} + C \quad \text{for } 0 < x < 2\pi.$$

The constant C can be found by letting $x = \pi$, which gives

$$C = \sum_{k=1}^{\infty} \frac{(-1)^{k-1}}{k^2}.$$

We thus have the equality

$$-\sum_{k=1}^{\infty} \frac{\cos kx}{k^2} = -\frac{(\pi - x)^2}{4} + \sum_{k=1}^{\infty} \frac{(-1)^{k-1}}{k^2} \quad \text{for } 0 < x < 2\pi,$$

that is, after changing signs

$$\frac{\cos x}{1^2} + \frac{\cos 2x}{2^2} + \frac{\cos 3x}{3^2} + \cdots = \frac{(\pi - x)^2}{4} - \left(\frac{1}{1^2} - \frac{1}{2^2} + \frac{1}{3^2} - \cdots\right)$$

for $0 < x < 2\pi$. But the series on the left converges uniformly and its sum is a

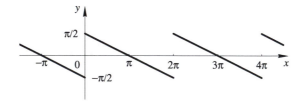

Fig. 7.16 Graph of the sum $\sum_{k=1}^{\infty} \frac{\sin kx}{k}$.

function continuous for all $x \in \mathbb{R}$. Thus, at $x = 0$,

(3) $$\frac{1}{1^2} + \frac{1}{2^2} + \frac{1}{3^2} + \cdots = \frac{\pi^2}{4} - \left(\frac{1}{1^2} - \frac{1}{2^2} + \frac{1}{3^2} - \cdots \right).$$

Hence we easily find

$$\frac{1}{1^2} + \frac{1}{3^2} + \frac{1}{5^2} + \cdots = \frac{\pi^2}{8}.$$

This equality was found by Euler. Using it we may write

$$\frac{1}{1^2} + \frac{1}{2^2} + \frac{1}{3^2} + \cdots = \left(\frac{1}{1^2} + \frac{1}{3^2} + \frac{1}{5^2} + \cdots \right) + \left(\frac{1}{2^2} + \frac{1}{4^2} + \frac{1}{6^2} + \cdots \right)$$

$$= \frac{\pi^2}{8} + \frac{1}{4} \left(\frac{1}{1^2} + \frac{1}{2^2} + \frac{1}{3^2} + \cdots \right)$$

and hence

(4) $$\frac{1}{1^2} + \frac{1}{2^2} + \frac{1}{3^2} + \cdots = \frac{\pi^2}{6}.$$

This equality was also given by Euler.
From (3) and (4) we easily obtain

$$\frac{1}{1^2} - \frac{1}{2^2} + \frac{1}{3^2} - \cdots = \frac{\pi^2}{12}.$$

Starting from the series $\left\{ \sum_{k=1}^{n} \cos(kx)/k \right\}$ and using a similar method we can obtain

(5) $$\frac{1}{1^4} + \frac{1}{2^4} + \frac{1}{3^4} + \cdots = \frac{\pi^4}{90}.$$

Comparing (4) and (5) we are anxious to find for what number α would we have

$$\frac{1}{1^3} + \frac{1}{2^3} + \frac{1}{3^3} + \cdots = \frac{\pi^3}{\alpha}.$$

The answer is quite unexpected: α is probably an irrational number, but we do not know whether this has ever been proved. The approximate value of α is 25.79435017.

Exercises 7.21.

1. Prove the following:

    (a) $$\frac{\sin x}{1} + \frac{\sin 2x}{2} + \frac{\sin 3x}{3} + \cdots = \frac{x}{2} \quad \text{for } -\pi < x < \pi.$$

(b) $\dfrac{\sin x}{1} + \dfrac{\sin 3x}{3} + \dfrac{\sin 5x}{5} + \cdots = \begin{cases} \dfrac{-\pi}{4} & \text{for } -\pi < x < 0; \\ \dfrac{\pi}{4} & \text{for } 0 < x < \pi. \end{cases}$

(c) $\dfrac{\sin x}{1^3} + \dfrac{\sin 2x}{2^3} + \dfrac{\sin 3x}{3^3} + \cdots = \dfrac{x(\pi - x)(2\pi - x)}{12}$ for $0 < x < 2\pi$.

2. Prove the following:

(a) $\dfrac{1}{1^3} - \dfrac{1}{3^3} + \dfrac{1}{5^3} - \cdots = \dfrac{\pi^3}{32}$.

(b) $\dfrac{1}{1^4} + \dfrac{1}{2^4} + \dfrac{1}{3^4} + \cdots = \dfrac{\pi^4}{90}$.

3. Provide a detailed proof for (2).

7.22 GENERAL FOURIER SERIES

The following theorem gives us the general form of coefficients of the Fourier series of a periodic function. Since all the functions involved are periodic with period 2π one can think of them as functions defined only on $[0, 2\pi]$.

7.22.1 Theorem *If*

(1) $\qquad f(x) = \dfrac{1}{2}a_0 + \displaystyle\sum_{k=1}^{\infty} (a_k \cos kx + b_k \sin kx) \ i.n. \qquad on \ [0, 2\pi],$

then

(2)
$$a_0 = \dfrac{1}{\pi} \int_0^{2\pi} f(x)\,dx, \qquad a_k = \dfrac{1}{\pi} \int_0^{2\pi} f(x) \cos kx\,dx,$$

$$b_k = \dfrac{1}{\pi} \int_0^{2\pi} f(x) \sin kx\,dx, \qquad k = 1, 2, \ldots.$$

Proof. Note that (1) means that

$$\lim_{m \to \infty} \int_0^{2\pi} \left| \dfrac{1}{2}a_0 + \sum_{k=1}^{m} (a_k \cos kx + b_k \sin kx) - f(x) \right| dx = 0.$$

Thus f has to be an integrable function on $[0, 2\pi]$. Integrating (1) from 0 to 2π we get

$$\int_0^{2\pi} f(x)\,dx = \pi a_0,$$

so that the first equality in (2) holds. In order to find the remaining coefficients a_k we first multiply (1) by $\cos mx$. The series

$$\left\{ \frac{1}{2} a_0 \cos mx + \sum_{k=1}^{n} (a_k \cos mx \cos kx + b_k \cos mx \sin kx) \right\}$$

converges in norm and thus we also have

$$\int_0^{2\pi} \cos mx \, f(x) dx = \frac{1}{2} \int_0^{2\pi} a_0 \cos mx \, dx + \sum_{k=1}^{\infty} \int_0^{2\pi} (a_k \cos mx \cos kx + b_k \cos mx \sin kx) \, dx.$$

Since

$$\int_0^{2\pi} (\cos mx)^2 dx = \pi,$$

$$\int_0^{2\pi} \cos mx \cos kx \, dx = 0 \quad \text{for } m \neq k,$$

$$\int_0^{2\pi} \cos mx \sin kx \, dx = 0 \quad \text{for all } k, m \in \mathbb{N},$$

we get

$$\int_0^{2\pi} f(x) \cos mx \, dx = \pi a_m.$$

Similarly, multiplying (1) by $\sin mx$ and integrating from 0 to 2π we obtain

$$\int_0^{2\pi} f(x) \sin mx \, dx = \pi b_m.$$

Let f be a function of period 2π and integrable over $[0, 2\pi]$; then it is possible to find the Fourier series of f, namely

(3)
$$\left\{ \frac{1}{2} a_0 + \sum_{k=1}^{n} (a_k \cos kx + b_k \sin kx) \right\},$$

where the a_k and the b_k are defined by (2). But does series (3) converge to f? The answer depends on properties of f and what is meant by the word "converge." For example, if the derivative of f is continuous the series (3) converges to f uniformly. Continuity of f is not sufficient for uniform convergence of its Fourier series; in fact it does not even imply pointwise convergence. On the other hand, if f is continuous, then series (3) converges to f almost everywhere. This follows from the famous theorem proved in 1966 by the Swedish mathematician Lennart Carleson (b. 1928). Carleson proved that if f is a square integrable function on $[0, 2\pi]$ then its Fourier series converges to f almost everywhere. A function f on $[0, 2\pi]$ is called *square integrable* if it is integrable and the function f^2 is also integrable. Every function continuous on $[0, 2\pi]$ is square integrable. The theory of square integrable functions will not be developed in this book.

Exercises 7.22

1. In the proof of 7.22.1 we said that the series

    $$\left\{ \frac{1}{2}a_0 \cos mx + \sum_{k=1}^{n} (a_k \cos mx \cos kx + b_k \cos mx \sin kx) \right\}$$

    converges in norm. Justify that claim.

2. Define

    $$f(x) = \begin{cases} 0 & \text{if } 0 \le x < \pi \\ 1 & \text{if } \pi < x \le 2\pi \end{cases}$$

    Find the Fourier series of f and check for convergence.

3. Define

    $$f(x) = \begin{cases} x & \text{if } 0 \le x < \pi \\ 2\pi - x & \text{if } \pi < x \le 2\pi \end{cases}$$

    Find the Fourier series of f and check for convergence.

7.23 EULER'S CONSTANT

The series $\left\{ \sum_{k=1}^{n} 1/k \right\}$ is called the *harmonic series*. It was proved in Section 7.20 to be divergent. This means that the sequence of partial sums $\frac{1}{1} + \cdots + \frac{1}{n}$ tends to infinity. However, it can be proved that the sequence

$$c_n = \frac{1}{1} + \frac{1}{2} + \cdots + \frac{1}{n} - \ln n$$

is convergent. The limit $C = \lim_{n \to \infty} c_n$ is called *Euler's constant*.

One may write

$$c_n = \alpha_1 + \cdots + \alpha_{n-1} + \frac{1}{n}$$

where

$$\alpha_n = \frac{1}{n} + \ln n - \ln(n+1)$$

because the intermediate logarithms cancel one another.

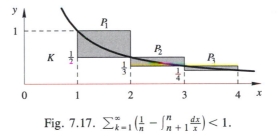

Fig. 7.17. $\sum_{k=1}^{\infty}\left(\frac{1}{n}-\int_{n+1}^{n}\frac{dx}{x}\right)<1$.

The curve in Fig. 7.17 is the graph of the function $f(x)=\frac{1}{x}$. On this curve lie vertices of the shaded rectangles $P_1, P_2, \ldots$. It is easily seen that the sum of the areas of all the rectangles is 1. In fact, were all the rectangles shifted horizontally to the square K, they would fill up the whole square. The part of the rectangle P_n that is above the curve has area

$$\frac{1}{n}-\int_{n}^{n+1}\frac{dx}{x}=\alpha_n.$$

Hence it follows that, for every $n \in \mathbb{N}$, $\alpha_n>0$ and the sum $\alpha_1+\cdots+\alpha_n$ is less than 1, because it is less than the sum of the areas of the rectangles P_n. This proves the convergence of the sequence $c_1, c_2, \ldots$. Moreover, we know that $\lim_{n\to\infty} c_n = C<1$. It is also easy to estimate C from below. To this aim it suffices to remark that the curve $y=1/x$ is concave-up and, consequently, the area of the upper part of the rectangle, lying above the curve, is greater than the area of half of the rectangle. Hence $C>1/2$, and thus

$$\tfrac{1}{2}<C<1.$$

In order to find a better estimate of C we consider the sequence

$$\gamma_n=c_n-\frac{1}{2n}+\frac{1}{12n^2}.$$

This sequence is also convergent to C. One can easily check that

$$\gamma_n-\gamma_{n+1}=\frac{1}{6}\int_{n}^{\infty}\frac{dx}{x^3(x+1)^3}.$$

Since

$$\frac{1}{x^3(x+1)^3}=\frac{1}{5}\frac{5x^4+10x^3+5x^2}{x^5(x+1)^5}<\frac{1}{5}\frac{(x+1)^5-x^5}{x^5(x+1)^5}=\frac{1}{5}\left(\frac{1}{x^5}-\frac{1}{(x+1)^5}\right),$$

we have

$$\frac{1}{6}\int_{n}^{\infty}\frac{dx}{x^3(x+1)^3}<\frac{1}{30}\int_{n}^{\infty}\left(\frac{1}{x^5}-\frac{1}{(x+1)^5}\right)dx=\frac{1}{30}\int_{n}^{n+1}\frac{dx}{x^5}$$

and

$$0 < \gamma_n - \gamma_{n+1} < \frac{1}{30} \int_n^{n+1} \frac{dx}{x^5}.$$

Hence

$$0 < \gamma_n - C < \frac{1}{30} \int_n^{\infty} \frac{dx}{x^5} = \frac{1}{120 n^4}.$$

Thus we have proved that by approximating C by

$$\gamma_n = \left(\frac{1}{1} + \frac{1}{2} + \cdots + \frac{1}{n} \right) - \frac{1}{2n} + \frac{1}{12 n^2} - \ln n$$

we make an error that does not exceed $1/120 n^4$. Using γ_{10}, for example, we make an error less than $1/1,200,000$. In the following calculations all numbers are rounded down.

$$\frac{1}{1} + \frac{1}{2} + \frac{1}{3} + \frac{1}{4} + \frac{1}{5} + \frac{1}{6} + \frac{1}{7} + \frac{1}{8} + \frac{1}{20} = \frac{21}{8}$$

$$
\begin{array}{rl}
21/8 = & 2.625 \\
1/7 \approx & 0.142857 \\
1/9 \approx & 0.111111 \\
1/1200 \approx & 0.000833 \\
\hline
 & 2.879801 \\
-\ln 10 \approx & -2.302585 \\
\hline
 & 0.577216
\end{array}
$$

In view of the inequality $1/1,200,000 < 1/10^6$, we have

$$0.577216 - \frac{1}{10^6} < C < 0.577216.$$

This shows that our calculations have given five reliable digits.
    A more exact value is

$$C \approx 0.5772156649.$$

Exercises 7.23.

1.   In this section we used the equality

$$\gamma_n - \gamma_{n+1} = \frac{1}{6} \int_n^{\infty} \frac{dx}{x^3 (x + 1)^3}.$$

Verify it.

2. Explain in detail why $\gamma_n - \gamma_{n+1} < \frac{1}{30} \int_n^{n+1} x^{-5}\, dx$, implies that
    $0 < \gamma_n - C < \frac{1}{30} \int_n^{\infty} x^{-5}\, dx$.

7.24 FUNCTIONS OF TWO REAL VARIABLES

This book is, in principle, concerned with functions of a single real variable. However, we occasionally used functions of two variables, for example in the Cauchy formula (Section 7.18). The function that to a number x assigns the value x^α is actually a function of two variables: x and the parameter α. From the mathematical point of view there is no reason to make a distinction between a function of one variable with a parameter and a function of two variables. In either case we mean a function that assigns to a pair of numbers x and y a real number $f(x, y)$. From now on we shall restrict ourselves to continuous functions for which x ranges over an interval I and y over an interval J, so that all pairs (x, y) represent points in the rectangle $I \times J = \{(x, y) : x \in I \text{ and } y \in J\}$. This will be sufficient for considering Euler's B and Γ functions (read beta and gamma) and for establishing the connection between them.

7.24.1 Definition   A function f defined on a rectangle is said to be *continuous* at a point (x_0, y_0) of the rectangle, if for every $\varepsilon > 0$ there exists a number $\delta > 0$ such that

$$|f(x, y) - f(x_0, y_0)| < \varepsilon$$

for all points (x, y) of the rectangle such that $|x - x_0| < \delta$ and $|y - y_0| < \delta$. A function which is continuous at every point of a rectangle is called continuous in that rectangle.

If the variable y of a function $f(x, y)$ is kept fixed at the value $y = y_0$, then we get the function of a single variable $f_1(x) = f(x, y_0)$. Similarly, if the variable x is kept fixed at $x = x_0$, we have the function $f_2(y) = f(x_0, y)$ of the single variable y. Clearly if f is continuous then both functions f_1 and f_2 are continuous. It might seem also that, conversely, if f_1 and f_2 are continuous, then f also is continuous. This, however, is not true, as the following example shows.

7.24.2 Example   Let $f(x, y) = xy/(x^2 + y^2)$. Then the function is defined everywhere except at the origin $(0, 0)$. If we additionally define $f(0, 0) = 0$, then the function becomes continuous with respect to each variable separately; that is, the functions f_1 and f_2 are continuous. On the other hand, for every $\delta > 0$ we have $f(\delta, \delta) = \frac{1}{2}$. So $|f(\delta, \delta) - f(0, 0)| = \frac{1}{2}$, which implies that f is not continuous at the origin $(0, 0)$.

7.24.3 Theorem   *A function f that is continuous on a bounded closed rectangle $[a, b] \times [c, d]$ is bounded, that is, there is a constant M such that $|f(x, y)| \leq M$ for all $(x, y) \in [a, b] \times [c, d]$.*

Proof. Assume, conversely, that f is not bounded. This means that for every $n \in \mathbb{N}$ there is a point $(x_n, y_n) \in [a, b] \times [c, d]$ such that $|f(x_n, y_n)| > n$. Since $a \le x_n \le b$ for all $n \in \mathbb{N}$, there exists, by the Bolzano–Weierstrass theorem, a subsequence $\{x_{p_n}\}$ that converges to some $x_0 \in [a, b]$. Now we consider the sequence $\{y_{p_n}\}$. Since $c \le y_{p_n} \le d$ for all $n \in \mathbb{N}$, there exists a subsequence $\{y_{q_n}\}$ of $\{y_{p_n}\}$ that converges to some $y_0 \in [c, d]$. It is easily seen that (x_0, y_0) is a point where f is not continuous. Indeed, for every $\delta > 0$ there is $n \in \mathbb{N}$ such that $|x_{q_n} - x_0| < \delta$, $|y_{q_n} - y_0| < \delta$ and at the same time $|f(x_{q_n}, y_{q_n}) - f(x_0, y_0)| > 1$.

7.24.4 Theorem   *If a function f is continuous on a bounded closed rectangle $[a, b] \times [c, d]$, then the function*

$$(1) \qquad F(x, y) = \int_c^y f(x, u)\,du$$

is continuous on the same rectangle.

Proof.   We have

$$
\left| F(x, y) - F(x_0, y_0) \right| = \left| \int_c^y f(x, u)\,du - \int_c^{y_0} f(x_0, u)\,du \right|
$$

$$
= \left| \int_c^y f(x, u)\,du - \int_c^y f(x_0, u)\,du + \int_{y_0}^y f(x_0, u)\,du \right|
$$

$$
\le \left| \int_c^y f(x, u) - f(x_0, u)\,du \right| + \left| \int_{y_0}^y f(x_0, u)\,du \right|
$$

Since f is continuous, for every $\varepsilon > 0$ there is some $\delta > 0$ such that $|x - x_0| < \delta$ implies that $\left| \int_c^y f(x, u) - f(x_0, u)\,du \right| < \frac{1}{2}\varepsilon$, and such that $|y - y_0| < \delta$ implies that $\left| \int_{y_0}^y f(x_0, u)\,du \right| < \frac{1}{2}\varepsilon$. Hence

$$\left| F(x, y) - F(x_0, y_0) \right| < \varepsilon \quad \text{for } |x - x_0| < \delta \quad \text{and} \quad |y - y_0| < \delta,$$

which proves the continuity of F.

Since the function

$$F(x, d) = \int_c^d f(x, y)\,dy$$

is continuous on $[a, b]$, it is integrable over that interval, the integral being given by

$$\int_a^b F(x, d)\,dx = \int_a^b \left(\int_c^d f(x, y)\,dy \right) dx.$$

To avoid the inconvenient parentheses this iterated integral is usually written in the form

$$(2) \qquad \int_a^b dx \int_c^d f(x, y)dy.$$

Similarly we can consider the integral

$$(3) \qquad \int_c^d dy \int_a^b f(x, y)dx.$$

Integrals (2) and (3) differ only by the order of integration. We are going to show that they are equal. To this end we shall first prove a theorem about differentiation under the integral sign.

When considering derivatives of functions of two variables it is always necessary to indicate the variable with respect to which the differentiation is performed. We shall use the symbols

$$D_x f = \lim_{h \to 0} \frac{f(x + h, y) - f(x, y)}{h}, \qquad D_y f = \lim_{k \to 0} \frac{f(x, y + k) - f(x, y)}{k}.$$

These derivatives are called *partial derivatives*.

7.24.5. Theorem   *If a function F is continuous on a rectangle $[a, b] \times [c, d]$ and $D_y F$ is continuous and bounded on $[a, b] \times (c, d)$, then the function*

$$I(y) = \int_a^b F(x, y)dx$$

is differentiable on (c, d) and the derivative is given by

$$I'(y) = D_y \left(\int_a^b F(x, y)dx \right) = \int_a^b D_y F(x, y)dx \quad \text{for all } y \in (c, d).$$

Proof.   Let $y_0 \in (c, d)$. Then

$$\left| \frac{I(y_0 + k) - I(y_0)}{k} - \int_a^b D_y F(x, y_0)dx \right|$$

$$\leq \int_a^b \left| \frac{F(x, y_0 + k) - f(x, y_0)}{k} - D_y F(x, y_0) \right| dx$$

$$= \int_a^b \left| D_y F(x, y_0 + \xi) - D_y F(x, y_0) \right| dx,$$

with $0 < \xi < k$; ξ depends on both x and k. However, if $k \to 0$, then also $\xi \to 0$ and $D_y F(x, y_0 + \xi) \to D_y F(x, y_0)$ everywhere in (a, b), because of the continuity of $D_y F$. Since $D_y F$ is bounded, the last integral tends to 0, by the dominated convergence theorem.

7.24.6 Theorem *If a function f is continuous on a rectangle $[a, b] \times [c, d]$, then*

(4)
$$\int_a^b dx \int_c^d f(x, y)dy = \int_c^d dy \int_a^b f(x, y)dx.$$

Proof. Consider the functions

$$\Phi(y) = \int_a^b dx \int_c^y f(x, u)du \quad \text{and} \quad \Psi(y) = \int_c^y du \int_a^b f(x, u)dx, \qquad y \in (c, d).$$

Since f is continuous and bounded on $[a, b] \times [c, d]$, so is the function

$$F(x, y) = \int_c^y f(x, u)du,$$

by 7.24.4. We can apply Theorem 7.24.5 to Φ to get

$$\Phi'(u) = \int_a^b D_y F(x, u)dx = \int_a^b f(x, u)dx.$$

On the other hand,

$$\Psi'(u) = \int_a^b f(x, u)dx.$$

Since Φ and Ψ have the same derivative, we have $\Phi = \Psi + \gamma$ for some constant γ. From the definitions of Φ and Ψ we clearly have $\Phi(c) = \Psi(c)$. Therefore, $\Phi(u) = \Psi(u)$ for every $u \in [c, d]$. In particular $\Phi(d) = \Psi(d)$, which is the required formula (4).

The conditions in 7.24.6 can be relaxed:

7.24.7 Theorem *If a function f is continuous on an open rectangle $(a, b) \times (c, d)$ and one of the iterated integrals*

$$\int_a^b dx \int_c^d |f(x, y)|dy \quad or \quad \int_c^d dy \int_a^b |f(x, y)|dx$$

exists, then both integrals in the equality

$$(5) \qquad \int_a^b dx \int_c^d f(x,y)dy = \int_c^d dy \int_a^b f(x,y)dx$$

exist and the equality holds.

Proof.   First we shall prove that if one of the integrals

$$\int_a^b dx \int_c^d |f(x,y)|dy \quad \text{or} \quad \int_c^d dy \int_a^b |f(x,y)|dx$$

exists then the other exists, too. In fact, suppose that the integral

$$\int_a^b dx \int_c^d |f(x,y)|dy$$

exists, that is, the integral $F(x) = \int_c^d |f(x,y)|dy$ exists almost everywhere in (a,b) and the function F is integrable over (a,b). Let $[\alpha_n, \beta_n]$, $n = 1, 2, \ldots$, be subintervals of (a,b) such that

$$[\alpha_n, \beta_n] \subset [\alpha_{n+1}, \beta_{n+1}] \subset (a,b), \qquad \alpha_n \to a, \quad \text{and} \quad \beta_n \to b.$$

Similarly, let $[\gamma_n, \delta_n]$, $n = 1, 2, \ldots$, be subintervals of (c,d) such that

$$[\gamma_n, \delta_n] \subset [\gamma_{n+1}, \delta_{n+1}] \subset (c,d), \qquad \gamma_n \to c, \quad \text{and} \quad \delta_n \to d.$$

Then

$$(6) \qquad \int_{\alpha_m}^{\beta_m} dx \int_{\gamma_n}^{\delta_n} f(x,y)dy = \int_{\gamma_n}^{\delta_n} dy \int_{\alpha_m}^{\beta_m} f(x,y)dx, \qquad m, n \in \mathbb{N},$$

according to the preceding theorem. For $m, n \in \mathbb{N}$, define the functions

$$f_{mn}(x,y) = \begin{cases} f(x,y) & \text{for } (x,y) \in [\alpha_m, \beta_m] \times [\gamma_n, \delta_n] \\ 0 & \text{elsewhere,} \end{cases}$$

and

$$f_n(x,y) = \begin{cases} f(x,y) & \text{for } (x,y) \in (a,b) \times [\gamma_n, \delta_n], \\ 0 & \text{elsewhere.} \end{cases}$$

For fixed $n \in \mathbb{N}$, the integrals $\int_a^b |f_{mn}(x,y)|dx$ form an increasing sequence of integrable functions on (c,d). Moreover, for every $y \in [c,d]$,

$$(7) \qquad \int_a^b |f_{mn}(x,y)|dx \to \int_a^b |f_n(x,y)|dx \quad \text{as } m \to \infty,$$

and

$$\int_c^d dy \int_a^b |f_{mn}(x,y)|dx = \int_a^b dx \int_c^d |f_{mn}(x,y)|dy \le \int_a^b dx \int_c^d |f(x,y)|dy < \infty.$$

Therefore, by the monotone convergence theorem, the sequence $\{\int_a^b |f_{mn}(x,y)|\}$ converges in norm and hence the limit $\int_a^b |f_n(x,y)|dx$ is an integrable function. Similarly, the integrals $\int_a^b |f_n(x,y)|dx$ form an increasing sequence of integrable functions on (c,d); for every $y \in [c,d]$

$$(8) \qquad\qquad \int_a^b |f_n(x,y)|dx \to \int_a^b |f(x,y)|dx \quad \text{as } n \to \infty,$$

and

$$\int_c^d dy \int_a^b |f_n(x,y)|dx \le \int_a^b dx \int_c^d |f(x,y)|dy < \infty.$$

Therefore, by the monotone convergence theorem, the sequence $\{\int_a^b |f_{mn}(x,y)|\}$ converges in norm as $n \to \infty$, and the limit $\int_a^b |f(x,y)|dx$ is integrable over (c,d), which implies that the integral $\int_c^d dy \int_a^b |f(x,y)|dx$ exists. Note that

$$\int_a^b dx \int_c^d |f_{mn}(x,y)|dy = \int_c^d dy \int_a^b |f_{mn}(x,y)|dx, \qquad m,n \in \mathbb{N},$$

implies that

$$\int_a^b dx \int_c^d |f(x,y)|dy = \int_c^d dy \int_a^b |f(x,y)|dx.$$

Now, since

$$\left| \int_c^d f_{mn}(x,y)dy \right| \le \int_c^d |f(x,y)|dy \quad \text{and} \quad \left| \int_a^b f_{mn}(x,y)dx \right| \le \int_a^b |f(x,y)|dx$$

for all $m,n \in \mathbb{N}$, we have

$$\lim_{n \to \infty} \lim_{m \to \infty} \int_a^b dx \int_c^d f_{mn}(x,y)dy = \int_a^b dx \int_c^d f(x,y)dy$$

and

$$\lim_{n\to\infty}\lim_{m\to\infty}\int_c^d dy\int_a^b f_{mn}(x,y)dx = \int_c^d dy\int_a^b f(x,y)dx,$$

by the dominated convergence theorem. Moreover, since

$$\int_a^b dx\int_c^d f_{mn}(x,y)dy = \int_c^d dy\int_a^b f_{mn}(x,y)dx$$

for all $m,n\in\mathbb{N}$, we conclude

$$\int_a^b dx\int_c^d f(x,y)dy = \int_c^d dy\int_a^b f(x,y)dx.$$

Theorems 7.24.6 and 7.24.7 are similar. However, in 7.24.7 the integrand may be unbounded, which increases the applicability of the theorem. The proof is much more complicated because the existence of the iterated integrals has to be proven, while in the first theorem it follows immediately from the fact that the function is continuous and the domain of integration bounded.

Theorem 7.24.7 is a weak version of the famous Fubini theorem (Guido Fubini, 1879–1943). In that theorem the condition of the continuity of f is replaced by integrability over the rectangle $(a,b)\times(c,d)$, bounded or unbounded.

Exercises 7.24

1. Let $f(x,y) = x^2y/(x^2+y^2)$ if $x\neq 0$ or $y\neq 0$, and $f(x,y) = 0$ if $x = y = 0$. Prove that f is continuous at $(0,0)$.

2. Let f and g be continuous functions at (x_0,y_0). Prove the following:
   (a) $f+g$ is continuous at (x_0,y_0).
   (b) λf is continuous at (x_0,y_0) for any $\lambda\in\mathbb{R}$.
   (c) fg is continuous at (x_0,y_0).
   (d) f/g is continuous at (x_0,y_0) provided $g(x_0,y_0)\neq 0$.

3. Let f be a continuous function on $[a,b]\times[c,d]$ and let $x_0\in(a,b)$. Prove that $f(x,y)$ converges to $f(x_0,y)$ as $x\to x_0$ uniformly on $[c,d]$, that is, for every $\varepsilon>0$ there exists $\delta>0$ such that for all $y\in[b,c]$, $|f(x,y)-f(x_0,y)|<\varepsilon$ whenever $|x-x_0|<\delta$.

4. Let f be a continuous function on $[a,b]\times[c,d]$ and let $x_0\in(a,b)$. Prove that for every $\varepsilon>0$ there exists $\delta>0$ such that for all $y\in[b,c]$, $\int_c^y |f(x,y)-f(x_0,y)|<\varepsilon$ whenever $|x-x_0|<\delta$.

5. Justify (7) and (8).

7.25 EULER'S GAMMA AND BETA FUNCTIONS

By *Euler's gamma function* we mean

(1)
$$\Gamma(p) = \int_0^\infty x^{p-1} e^{-x} dx, \qquad p > 0$$

(In Example 7.17.3 we proved that the integral exists for every $p > 0$.)
We have

$$\Gamma(1) = \int_0^\infty e^{-x} dx = 1.$$

Integration by parts

$$\int_0^\infty x^{p-1} e^{-x} dx = \left[\frac{1}{p} x^p e^{-x} \right]_0^\infty + \frac{1}{p} \int_0^\infty x^p e^{-x} dx = \frac{1}{p} \int_0^\infty x^p e^{-x} dx$$

leads to the following important result

$$p\Gamma(p) = \Gamma(p + 1),$$

and hence, by induction,

$$\Gamma(p + 1) = p!$$

for all natural numbers p. We see that the gamma function gives a continuous extension to all positive real numbers of the factorial $p!$ defined only for natural numbers.

Euler's beta function is defined by

(2)
$$B(p, q) = \int_0^1 x^{p-1} (1 - x)^{q-1} dx, \qquad p > 0, \qquad q > 0.$$

Here the integrand is continuous on $[\varepsilon, 1 - \varepsilon]$ for every $0 < \varepsilon < \frac{1}{2}$ and is bounded on $(0, \frac{1}{2}]$ by $x^{p-1} k_q$ and on $[\frac{1}{2}, 1)$ by $m_p (1 - x)^{q-1}$, where k_q and m_p are properly chosen numbers depending on q and p, respectively.

The beta function can also be defined by

(3)
$$B(p, q) = \int_0^\infty \frac{t^{p-1}}{(1 + t)^{p+q}} dt,$$

which is obtained from (2) by the substitution $x = t/1 + t$. In fact, assuming

$$f(x) = x^{p-1}(1 - x)^{q-1} \quad \text{and} \quad g(t) = \frac{t}{1 + t},$$

we have

$$f(g(t))g'(t) = \frac{t^{p-1}}{(1+t)^{p-1}} \frac{1}{(1+t)^{q-1}} \frac{1}{(1+t)^2},$$

$g(0) = 0$, and $g(\infty) = 1$, and formula (3) follows from (2).

7.25.1 Theorem

$$B(p, q) = \frac{\Gamma(p)\Gamma(q)}{\Gamma(p+q)} \quad for\ all \quad p > 0 \quad and \quad q > 0.$$

Proof.   Substituting $x = ty$ in (1) we obtain

$$\frac{\Gamma(p)}{t^p} = \int_0^\infty y^{p-1}e^{-ty}\,dy, \qquad t > 0.$$

Writing now $p + q$ instead of p and $1 + t$ instead of t we get

$$\frac{\Gamma(p+q)}{(1+t)^{p+q}} = \int_0^\infty y^{p+q-1}e^{-(1+t)y}\,dy.$$

Multiplying both sides by t^{p-1} and integrating with respect to t from 0 to ∞ we have

$$\Gamma(p+q)\int_0^\infty \frac{t^{p-1}}{(1+t)^{p+q}}\,dt = \int_0^\infty t^{p-1}dt \int_0^\infty y^{p+q-1}e^{-(1+t)y}\,dy.$$

The integral on the left side equals $B(p, q)$. On the right side we interchange the order of integration (Theorem 7.24.7):

$$\Gamma(p+q)B(p, q) = \int_0^\infty y^{p+q-1}e^{-y}\,dy \int_0^\infty t^{p-1}e^{-ty}\,dt$$

$$= \Gamma(p)\int_0^\infty y^{q-1}e^{-y}\,dy.$$

Hence formula 7.25.1 follows.

This beautiful proof is due to the German mathematician Peter Gustav Lejeune Dirichlet (1805–1859).

From 7.25.1 we have $B(\frac{1}{2}, \frac{1}{2}) = \Gamma(\frac{1}{2})\Gamma(\frac{1}{2})$. But

$$B(\frac{1}{2}, \frac{1}{2}) = \int_0^1 \frac{dt}{\sqrt{t(1-t)}} = \int_0^1 \frac{dt}{\sqrt{\frac{1}{4} - (t-\frac{1}{2})^2}} = \left[\arcsin\left[2\left(t - \frac{1}{2}\right)\right]\right]_0^1 = \frac{\pi}{2} + \frac{\pi}{2} = \pi.$$

Thus $\Gamma(\frac{1}{2})\Gamma(\frac{1}{2}) = \pi$, whence $\Gamma(\frac{1}{2}) = \sqrt{\pi}$, that is,

$$\int_0^\infty \frac{e^{-t}}{\sqrt{t}}\,dt = \sqrt{\pi}.$$

Substituting $t = x^2$ we get $2\int_0^\infty e^{-x^2}\,dx = \sqrt{\pi}$ and hence

$$\int_{-\infty}^\infty e^{-x^2}\,dx = \sqrt{\pi}.$$

This integral is called the *Euler-Poisson integral* (Siméon Denis Poisson, 1781–1840).

HINTS AND ANSWERS FOR SELECTED EXERCISES

EXERCISES 1.1

1. Check I.
2. Check II.
3. Check II or III.
4. Check II or III.

EXERCISES 1.2

2. Consider a set $X = \{x, y, z\}$ with $x < y$, $y < z$, and $z < x$ being the only inequalities in X. Then I and III are satisfied, but 1.2.4 is no longer true.

EXERCISES 1.3

1. Yes

EXERCISES 1.4

1. To prove that IV does not follow from I, II, and III take $X = \{1, 2, 3\}$ with $1 < 2 < 3$.
2. Explain carefully the meaning of "axioms I–IV fully characterize natural numbers."
4. Find the place in the proof of 1.4.1 where IV is used.

EXERCISES 1.5

1. Use 1.5.3 and 1.5.4.
2. Use 1.5.3.
3. Use 1.5.1 and 1.5.6.

EXERCISES 1.6

1. Use 1.6.1.
2. Use 1.5.1, 1.5.6, and 1.6.1.
3. Use 1.5.1, 1.5.6, and 1.6.1.
4. Use 1.5.1, 1.5.6, and 1.6.1.

EXERCISES 2.1

1. Yes
2. Yes
3. Yes

EXERCISES 2.3

1. Yes
2. Consider the equation $2x = 3$.
3. Yes

EXERCISES 2.5

2. Yes
3. Yes

EXERCISES 2.6

3. No
4. No
7. Yes
8. No
9. Yes

EXERCISES 2.7

4.   Prove, by induction, that $0 \leq a_n^2 \leq 2$ for all $n \in \mathbb{N}$.

EXERCISES 2.9

2.   No
3.   No
4.   Yes
5.   Yes
6.   False

EXERCISES 2.12

13.   Use Theorem 2.12.5 instead of repeating the proof.
15.   False
16.   Prove, by induction, that $a_n^2 \leq 2$ and $a_n \leq a_{n+1}$, and then use Theorem 2.12.5.

EXERCISES 2.14

1.   Yes
2.   No
3.   No
4.   No
5.   Yes. Use Theorem 2.14.1.
7.   Prove that every nonempty interval contains uncountably many irrational numbers.

EXERCISES 3.1

1. $t = \sqrt{h/16}$.
2.   (a) $[1, \infty)$
     (b) $[0, \infty)$
     (c) $(0, 1]$
     (d) $\mathbb{R}$
     (e) $\mathbb{R}$
     (f)   The set of all integers
     (g)   The set of all natural numbers greater than 6

3. (a) Yes
    (b) Yes
    (c) No
    (d) Yes
    (e) Yes

EXERCISES 3.2

1. (a)

(b)

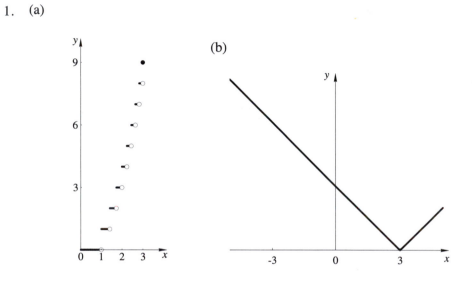

(c)

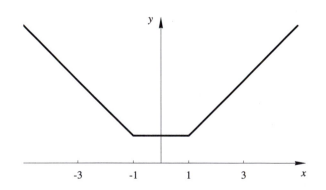

(d)

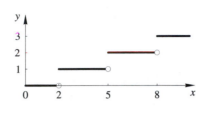

EXERCISES 3.3

1. (a)

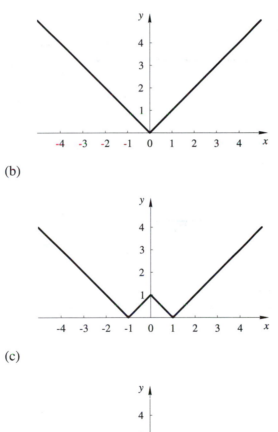

(b)

(c)

(d)

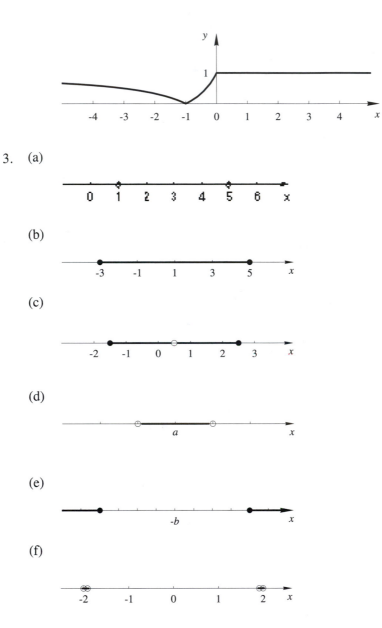

3. (a)

(b)

(c)

(d)

(e)

(f)

4. Consider cases $x < y$, $x = y$, and $x > y$.

EXERCISES 3.4

4. Note that $0 \leq |f(x)| \leq |x|$. Use this to show that $\lim_{x \to 0} f(x) = 0$.

EXERCISES 3.5

1. (a) 2/3
    (b) -1
    (c) 0
    (d) 0
2. (a) $2x$
    (b) $-2/x^3$
    (c) -1
    (d) 1

EXERCISES 3.6

1. Show that $0 \le 1 - x\,[1/x] \le x$ for $x > 0$.
2. Show that $\left| x\,[2/x] + x/2 - 2 \right| \le x/2$ for $x > 0$.
3. Note that $1 - x^3 \le f(x) \le 1 + x^2$ for $|x| \le 1$.
5. Note that $\lim_{x \to x_0} f(x) = \lim_{x \to 0} f(x + x_0)$.
6. False

EXERCISES 3.7

1. $f(2) = 5$.
4. Modify the function in Exercise 3.4.3.
6. Compare to Exercise 3.3.4.
7. Compare to Exercise 3.4.3.
8. Compare to Exercise 3.4.4.
9. Note that if a sequence of rational numbers p_n/q_n converges to a number x and $p_n/q_n \ne x$, then $q_n \to \infty$.
10. Modify the function in Exercise 3.7.8 to get a function continuous at a and a function continuous at b, and then multiply them.
12. Use the fact that every number is a limit of a sequence of rational numbers.
13. Compare to Exercise 3.7.12.
14. Prove first that $f(x) = f(1)x$ for all rational numbers x and then use continuity of f.

EXERCISES 3.8

2. Use induction and Theorem 3.7.3.
3. Use Exercise 3.8.2 and Theorem 3.7.3.
4. Use induction and Theorem 3.7.3.

EXERCISES 3.9

2. Compare to the example following Theorem 3.9.3.
3. Consider the function $h(x) = x - f(x)$.
4. Consider the function $h(x) = f(x) - g(x)$.
5. Use Theorem 3.9.4.
6. Show that if the function is not constant then its range contains an interval.

EXERCISES 3.10

1. Use a discontinuous function.
2. False
3. Note that there exists $M > 0$ such that $|f(x)g(x)| \leq M|f(x)|$ for all $x \in [x_0 - \delta, x_0 + \delta]$.
6. False
7. True
8. False

EXERCISES 3.11

2. Use a discontinuous function.
4. Use an open interval.
5. False
6. False
7. False
8. False

EXERCISES 3.12

2. Yes
3. (a) Yes
   (b) Yes
   (c) Yes

EXERCISES 3.13

1. 4
2. 6

EXERCISES 3.14

2. $y = 3x - 2$.
3. $(1/2, 1/4)$

EXERCISES 3.15

2. To prove differentiability at 0 use the fact that $0 \le f(x) \le x^2$ and hence

$$0 \le \left| \frac{f(h) - f(0)}{h} \right| \le |h|.$$

At other points check for continuity.
3. Yes

EXERCISES 3.16

2. Differentiate the equality $1 = f \cdot (1/f)$.
4. Yes
5. $(fgh)' = f'gh + fg'h + fgh'$

EXERCISES 3.17

1. Use induction and Theorems 3.16.1 and 3.17.1.
2. Yes

EXERCISES 3.18

2. Use Exercise 3.17.1 and Theorem 3.18.1.
3. Compare the denominators of f and f'.

EXERCISES 3.19

2. Find the minimum of $f(x) = x + 1/x$ for $x > 0$.
3. $p = -2$ and $q = 4$.

EXERCISES 3.20

1. Use $f(x) = |x|$.
2. The domain of f can consist of two disjoint intervals.
3. Use Theorem 3.20.1.
4. Use Theorem 3.20.1.
5. Consider the function $h(x) = g(x) - f(x)$.

EXERCISES 3.21

2. Let $x > 0$. Then

$$\frac{f(x)}{x} = \frac{f(x) - f(0)}{x - 0} = f'(\xi) < f'(x)$$

and consequently

$$g'(x) = \frac{f'(x)x - f(x)}{x^2} > 0.$$

3. False

EXERCISES 3.23

3. $\displaystyle\lim_{x\to\infty} \frac{a_n x^n + \cdots + a_1 x + a_0}{b_m x^m + \cdots + b_1 x + b_0} = \begin{cases} 0 & \text{if } m > n \\ a_n/b_m & \text{if } m = n \\ \infty & \text{if } m < n \text{ and } a_n b_m > 0 \\ -\infty & \text{if } m < n \text{ and } a_n b_m < 0 \end{cases}$

EXERCISES 4.1

2. Note that $0 \le 1/n! \le 1/n$.
4. (a) No
   (b) Yes
   (c) Yes
   (d) Yes
5. $f(x) = (a_{n+1} - a_n)(x - n) + a_n$ for $x \in [n, n+1]$.
6. False
10. False

EXERCISES 4.2

1. (a) Yes
   (b) Yes
   (c) No

2.   Compare with Exercise 2.12.16.

3.   Consider separately the sequence of even-numbered terms and odd-numbered terms. Compare with Exercise 4.2.2.

4.   Prove that the sequence $\{ (1 + 1/n)^n \}$ is increasing and bounded. (This is not easy.)

EXERCISES 4.3

1.   Note that $\{q^{n^2}\}$ is a subsequence of $\{q^n\}$.

2.   0

3.   0

4.   0

EXERCISES 4.4

1.   Note that if $|q| < 1$ then $|q^k| < 1$. Use Theorem 4.4.1 with q^k instead of q.

2.   Note that if $|q| < 1$ then $|-q| < 1$. Use Theorem 4.4.1 with $-q$ instead of q.

EXERCISES 4.5

1.   Note that $\frac{1}{k(k+1)} = \frac{1}{k} - \frac{1}{k+1}$.

3.   (a)   Divergent

     (b)   Divergent

     (c)   3/4

     (d)   5/12

     (e) $-8/17$

     (f)   45

4.   Note that

$$\left(\frac{1}{3} + \frac{1}{4}\right) \geq \frac{1}{2}, \qquad \left(\frac{1}{5} + \frac{1}{6} + \frac{1}{7} + \frac{1}{8}\right) \geq \frac{1}{2},$$

$$\left(\frac{1}{9} + \frac{1}{10} + \frac{1}{11} + \frac{1}{12} + \frac{1}{13} + \frac{1}{14} + \frac{1}{15} + \frac{1}{16}\right) \geq \frac{1}{2},$$

and so on.

5.   No

6.   Yes

EXERCISES 4.6

1.   (a)   Note that $\dfrac{1}{2^k + \sqrt{k}} \leq \dfrac{1}{2^k}$.

(b)   Note that $\left| \dfrac{(-1)^k}{k - 3^k} \right| \le \dfrac{1}{2^k}$.

3.   Use Theorem 4.6.1.

4.   Calculate partial sums.

5.   Use the fact that the series $\left\{ \sum_{k=1}^{n} 1/k \right\}$ diverges.

6.   False

7.   Note that $a_n = p_n - q_n$.

8.   Follow the construction described in Section 4.6.

12.   Note that

$$\sum_{n=1}^{\infty} \frac{n}{2^n} = \sum_{n=1}^{\infty} \frac{1}{2^n} + \sum_{n=2}^{\infty} \frac{1}{2^n} + \sum_{n=3}^{\infty} \frac{1}{2^n} + \cdots .$$

13.   Note that if $\lim_{n \to \infty} a_n / b_n = L \ne 0$ then $|a_n| \le |2L||b_n|$ and $|b_n| \le |2/L||a_n|$ for almost all $n \in \mathbb{N}$.

14.   Consider $\left\{ \sum_{k=1}^{n} (-1)^k / \sqrt{k} \right\}$ and $\left\{ \sum_{k=1}^{n} \left((-1)^k / \sqrt{k} + 1/k \right) \right\}$.

EXERCISES 4.9

2.   Substitute $z = x^m$.

3.   Modify the proof of Theorem 4.9.2.

EXERCISES 4.10

1. $f(x) = \begin{cases} 0 & \text{if } x \in [0, 1) \\ 1 & \text{if } x = 1 \end{cases}$

2.   (a) $f(x) = \begin{cases} 1 & \text{if } x \ne 0 \\ 0 & \text{if } x = 0 \end{cases}$

(b) $f(x) = \begin{cases} \dfrac{1 + x^2}{x} & \text{if } x \ne 0 \\ 0 & \text{if } x = 0 \end{cases}$

(c) $f(x) = x$

3.   Let $x \in (0, \infty)$ be fixed. Define $x_1 = \sqrt{x}$ and $x_{n+1} = \sqrt{x + x_n}$ for $n \ge 2$. Assuming that the series converges find its limit. Then, using induction, show that the sequence is bounded by that limit.

EXERCISES 4.11

1.   Use the definition of uniform convergence.

2.   Note that $\max_{x \in [a,b]} |g(x)| \le M$ if and only if $|g(x)| \le M$ for all $x \in [a, b]$.

3. (a), (b), and (c): Not uniformly convergent

4. False

5. True

7. Use the mean value theorem.

EXERCISES 4.12

1. Use the fact that $\sum_{k=0}^{\infty} x^k = 1/(1-x)$ for $x \in (-1, 1)$.

3. Consider the sequence of functions

$$f_n(x) = \begin{cases} \dfrac{1 - |2n(n+1)x - 2n - 1|}{n} & \text{if } \dfrac{1}{n+1} \leq x \leq \dfrac{1}{n} \\ 0 & \text{otherwise} \end{cases}$$

4. Note that the Weierstrass test cannot be used here. Use either Definition 4.11.1 or Exercise 4.11.1.

EXERCISES 4.13

2. Apply Theorem 4.13.1 to partial sums.

3. (a) Note that $\left(\sum_{k=1}^{\infty} x^k\right)' = \sum_{k=1}^{\infty} k x^{k-1}$.

   (b) Differentiate $\sum_{k=1}^{\infty} x^k$ twice.

4. Use Exercise 4.13.3.

EXERCISES 4.16

1. (a) $f'(g(x) + h(x))(g'(x) + h'(x))$
   (b) $f'(g(x)h(x))(g'(x)h(x) + g(x)h'(x))$
   (c) $f'(f(f(x)))f'(f(x))f'(x)$

2. (a) Yes
   (b) Yes
   (c) Yes

EXERCISES 4.17

1. Modify the argument presented in Section 4.17.

EXERCISES 4.20

1. $\sqrt[5]{e} \approx 1.221403$ and $1/\sqrt{e} \approx 0.606531$.

EXERCISES 4.21

1. It is possible.
3. False
4. False

EXERCISES 4.24

1. (a) $3x(2 \ln x + 1)$

    (b) $\dfrac{2x}{1 + x^2}$

    (c) $\dfrac{2}{(1 + x)(1 - x)}$

    (d) $e^x \ln x + \dfrac{e^x}{x}$

    (e) $\dfrac{1}{x \ln x}$

    (f) $\dfrac{-1}{e^x + 1}$

2. $\ln \frac{1}{2} \approx -0.69314718.$

EXERCISES 4.25

2. (a) $2^x \ln 2$

    (b) $\dfrac{1}{(1 + x) \ln 2}$

    (c) $\dfrac{-3^x \ln 3}{1 + 3^x}$

    (d) $2^{(2^x)} 2^x (\ln 2)^2$

EXERCISES 4.26

1. (a) $\dfrac{1}{x(1 + x)}$

    (b) $\dfrac{\frac{1}{\sqrt{1+x}} + \frac{1}{\sqrt{1-x}}}{\sqrt{1 + x} - \sqrt{1 - x}} - \dfrac{1}{x}$

    (c) $3^x \ln 3 + 3x^2 + x^x (\ln x + 1)$

    (d) $-2x^{-2x} (\ln x + 1)$

    (e) $\dfrac{8x^4 - 3x^2 + 4x + 6}{15x^3 \cdot \sqrt{x^2 - 1}}$

(f) $\dfrac{2}{(2 - \sqrt{x})^2 \cdot \sqrt{x}}$

(g) $\dfrac{2^{\sqrt{1+2x}}}{\sqrt{1+2x}} \ln 2$

(h) $(\log_2 x)^{(e^x)} e^x \left(\ln(\log_2 x) + \dfrac{1}{x \ln x} \right).$

2.   (a) $(-\infty, 1)$

(b) (e, ∞)

(c) $(-2, \sqrt{5} - 1)$

EXERCISES 4.28

1. $999^{998} < 998^{999}.$

EXERCISES 4.35

2.   Prove that $f(x) = \arctan x + \arctan(1/x)$ is a constant function for $x > 0$.

EXERCISES 4.36

1.   (a) $2(\tan \sqrt{x})^3 \dfrac{1 + (\tan \sqrt{x})^2}{\sqrt{x}}$

(b) $\dfrac{1}{\cos x}$

(c) $(\tan x)^{\sin x} \cos x \left(\ln(\tan x) + (\tan x)^2 + 1 \right)$

(d) $4x^4 \arctan x + \dfrac{x^4}{1 + x^2}$

(e) $\dfrac{x}{(x^2 - 1) \cdot \sqrt{1 - x^2}}$

2.   (a)   Yes

(b)   No

(c)   Yes

3.   (a)   1000

(b)   About e^{1000} (this number has 435 digits).

(c)   6

EXERCISES 4.38

2.   Use (5).

3. (a) 2

(b) 1

(c) $\frac{1}{\sqrt{3}}$

EXERCISES 4.39

1. $\arccos x = \frac{\pi}{2} - x - \frac{1}{6}x^3 - \frac{3}{40}x^5 - \frac{5}{112}x^7 - \cdots.$

2. $\ln\left(x + \sqrt{1 + x^2}\right) = x - \frac{1}{6}x^3 + \frac{3}{40}x^5 - \frac{5}{112}x^7 + \frac{35}{1152}x^9 - \cdots.$

3.
(a) $\dfrac{3 \cdot \sqrt{x}}{2 \cdot \sqrt{1 - x^3}}$

(b) $\dfrac{1}{2 \cdot \sqrt{x(1 - x)}}$

(c) $3x^2 \arcsin \dfrac{1}{x} - \dfrac{x^2}{\sqrt{x^2 - 1}}$

(d) $\dfrac{-\arcsin\sqrt{1 - x^2}}{x^2}$

EXERCISES 5.2

1. (a) 1

(b) $\sqrt{3}|a|$

(c) $\ln a - \ln b$

(d) 2

(e) $1/3$

(f) 1

EXERCISES 5.3

1. (a) 0

(b) 1

(c) 0

3. Use induction.

EXERCISES 5.4

1. (a) $1/2$

(b) $-\infty$

(c) -1

(d) 0

(e) 1

(f) e^m

(g) $1/e$

(h) $e^{1/3}$

EXERCISES 5.5

1.   (a) 1

   (b) $4/e$

EXERCISES 5.6

3.   Note that $|a_n|^{1/n} < \alpha < 1$ for all n greater than some n_0. Thus $|a_n| < \alpha^n$ for all $n > n_0$.

4.   (a), (b), (d), (e), and (f) Convergent; (c) Divergent

EXERCISES 5.7

1.   Use the expansion of $\sqrt{1+x}$ into a power series.

EXERCISES 6.2

1.   (a) $3 \cdot 2^x / \ln 2 + 2 \cdot 3^x / \ln 3$

   (b) $b \sin x - a \cos x$

   (c) $\arctan x + \dfrac{1}{\sqrt{2}} \arctan \dfrac{x}{\sqrt{2}} + \dfrac{1}{\sqrt{3}} \arctan \dfrac{x}{\sqrt{3}}$

   (d) $\ln\left(x + \sqrt{1+x^2}\right) + \arcsin x$

   (e) $\dfrac{e^{ax+b}}{a}$

   (f) $\dfrac{\sin 2x}{2}$

   (g) $\dfrac{-\cos(3x+1)}{3}$

   (h) $\dfrac{2}{\sqrt{7}} \arctan \dfrac{4x-1}{\sqrt{7}}$

(i) $\quad \frac{1}{\sqrt{2}} \ln\left(\frac{1}{\sqrt{2}} \sqrt{1 - x + 2x^2} + 4x - 1 \right)$

(j) $\quad \frac{1}{\sqrt{2}} \arcsin \dfrac{4x + 1}{3}$

EXERCISES 6.3

1. (a) $x \arctan x - \ln \sqrt{1 + x^2}$

(b) $x \arcsin x + \sqrt{1 - x^2}$

(c) $x \arccos x - \sqrt{1 - x^2}$

(d) $\frac{1}{2}(\arctan x)^2$

(e) $\frac{1}{2}(\arcsin x)^2$

(f) $(6x - x^3)\cos x + (3x^2 - 6)\sin x$

(g) $\frac{1}{2}e^x(\cos x + \sin x)$

(h) $\frac{1}{2}(2x - 1)\arctan(2x - 1) - \frac{1}{4}\ln\left(1 + (2x - 1)^2\right)$

(i) $\frac{2}{289}e^{2x}\left((16 - 17x)\cos\frac{x}{2} - (30 - 68x)\sin\frac{x}{2}\right)$

EXERCISES 6.4

1. (a) $x + \frac{1}{4}\ln\dfrac{x - 1}{x + 1} - \dfrac{1}{2}\arctan x$

(b) $\frac{1}{3}\ln\dfrac{x + 1}{\sqrt{x^2 - x + 1}} + \dfrac{1}{\sqrt{3}}\arctan\dfrac{2x - 1}{\sqrt{3}}$

(c) $\frac{1}{4}\ln\dfrac{x^2 + x + 1}{x^2 - x + 1} + \dfrac{1}{2\sqrt{3}}\left(\arctan\dfrac{2x + 1}{\sqrt{3}} + \arctan\dfrac{2x - 1}{\sqrt{3}}\right)$

(d) $\frac{1}{2}\left(\dfrac{x}{1 + x^2} + \arctan x\right)$

(e) $\dfrac{x - \frac{1}{2}}{x^2 + 2x + 2} + \arctan(x + 1)$

(f) $\ln x - \frac{1}{7}\ln(x^7 + 1)$

(g) $\frac{1}{5}\left(\dfrac{1}{x^5 + 1} + \ln\dfrac{x^5}{x^5 + 1}\right)$

(h) $-\dfrac{1}{9(x - 1)^9} - \dfrac{1}{4(x - 1)^8} - \dfrac{1}{7(x - 1)^7}$

EXERCISES 6.5

1. (a) $-\sqrt{1 - x^2}\left(\frac{1}{6}x^5 + \frac{5}{24}x^3 + \frac{5}{16}x\right) + \frac{5}{16}\arcsin x$

(b) $\frac{1}{2}\left(\dfrac{\sqrt{x^2+2x}}{(x+1)^2} + \arccos\dfrac{1}{x+1}\right)$

(c) $\dfrac{\sqrt{1+x^2}}{3}\left(\dfrac{2}{x} - \dfrac{1}{x^3}\right)$

(d) $-\dfrac{1}{4x}(2+x^3)^{1/3}\left(1 + \dfrac{x^3}{2(2+x^3)}\right)$

(e) $-\dfrac{2\left(1+x^{3/4}\right)^{2/3}}{\sqrt{x}}$

EXERCISES 6.6

1.   (a) $\frac{1}{2}\left(\sin^2 x - \dfrac{1}{\sin^2 x}\right) - 2\ln(\sin x)$

(b) $\dfrac{1}{3\sin x}\left(\dfrac{1}{\cos^3 x} + \dfrac{4}{\cos x} - 8\cos x\right)$

(c) $\frac{1}{\sqrt{2}}\left(\ln(\sin x + \sqrt{\sin 2x} + \cos x) + \arctan\dfrac{\sqrt{2}\tan x}{1-\tan x}\right)$

(d) $\frac{1}{24}\cos 6x - \frac{1}{16}\cos 4x - \frac{1}{8}\cos 2x$

(e) $2\sqrt{\tan x}$

(f) $3\sin\dfrac{x}{6} + \dfrac{3}{5}\sin\dfrac{5x}{6}$

(g) $\frac{1}{\sqrt{2}}\ln\left(\dfrac{\tan(x/2) - 1 + \sqrt{2}}{\tan(x/2) - 1 - \sqrt{2}}\right)$

(h) $\arctan\left(\tan\dfrac{x}{2} + 1\right)$

EXERCISES 7.2

1.   Verify Ⅰ and Ⅲ.
2.   Verify Ⅱ and Ⅲ.

EXERCISES 7.3

1.   (a), (b), and (d) Yes; (c) and (e) No
2.   No

EXERCISES 7.4

1.   (a) $\chi_{[-2,-1)} - \chi_{[-1,1)} + \chi_{[1,2)}$

2. False

3. (c) Note that $0 \leq f$ implies $0 \leq \int f$ and use part (a).

   (d) Note that $-|f| \leq f \leq |f|$ and use part (c).

4. Use Exercise 7.4.3(c).

6. Use Lemma 7.4.5.

7. Use Lemma 7.4.5.

8. False

EXERCISES 7.5

1. Check ⓘ and ⓘⓘ.

2. Check ⓘ and ⓘⓘ.

EXERCISES 7.6

1. Compare with Exercise 3.3.4.

3. No. Consider $\chi_{\{a\}}$.

4. First assume that $f \geq 0$, and use only bricks with positive coefficients.

EXERCISES 7.7

2. $f = \sum_{k=1}^{\infty} \chi_{\{p_n\}}$ where $\mathbb{Q} = \{p_1, p_2, \ldots\}$.

3. Compare with Exercise 7.7.2.

EXERCISES 7.8

3. Note that integrability of g is not assumed.

4. Use Exercise 7.8.3.

5. Note that $fg = g$ a.e.

EXERCISES 7.9

1. Note that $|f + g| \leq |f| + |g|$.

4. Find an expansion for f_h.

5. False

EXERCISES 7.10

3. False

EXERCISES 7.11

3. False

EXERCISES 7.13

1. Compare with Exercise 3.3.4. Use induction.

EXERCISES 7.14

1. False

EXERCISES 7.15

3. Note that $\min_{[a,b]} f(x)(b-a) \le \int_a^b f \le \max_{[a,b]} f(x)(b-a)$ and use Theorem 3.9.4.
4. Consider the function $h = f - g$ and use Exercise 7.15.3.
6. Differentiate both sides of $\int_a^x f = \int_x^b f$.

EXERCISES 7.16

1. (a) π
   (b) -1

EXERCISES 7.17

1. (a) $\ln \frac{4}{3}$
   (b) $\frac{1}{3} \ln \frac{5}{4}$
   (c) π
   (d) $\frac{1}{2} \left(\sqrt{2} + \ln \left(1 + \sqrt{2} \right) \right)$
   (e) 2
   (f) $2\sqrt{3}$

EXERCISES 7.20

1. (a) Converges
   (b) Diverges

EXERCISES 7.22

2. $\frac{1}{2} - \frac{2}{\pi}\cos x - \frac{2}{3\pi}\cos 3x - \frac{2}{5\pi}\cos 5x - \cdots$

3. $\frac{\pi}{2} - \frac{4}{\pi}\cos x - \frac{4}{9\pi}\cos 3x - \frac{4}{25\pi}\cos 5x - \cdots$

EXERCISES 7.24

3. Suppose that it is not true. Then there exist $(x_n, y_n) \in [a, b] \times [c, d]$ such that $x_n \to x_0$ and $|f(x_n, y_n) - f(x_0, y_n)| \geq \varepsilon$ for some $\varepsilon > 0$ and all $n \in \mathbb{N}$. Use the Bolzano–Weierstrass theorem.

4. Use Exercise 7.24.3.

INDEX